FORAMINIFERA OF THE ORINOCO-TRINIDAD-PARIA SHELF

REPORTS OF THE ORINOCO SHELF EXPEDITION

Volume IV

BY

C. W. DROOGER and J. P. H. KAASSCHIETER

With 4 textfigures, 5 plates and 41 maps

VERHANDELINGEN DER KONINKLIJKE NEDERLANDSE AKADEMIE VAN WETENSCHAPPEN, AFD. NATUURKUNDE

EERSTE REEKS, DEEL XXII

1958
N.V. NOORD-HOLLANDSCHE UITGEVERS MAATSCHAPPIJ
AMSTERDAM

Aangeboden oktober 1957
Aanvaard november 1957
Gepubliceerd december. 1958

CONTENTS

PREFACE

This report on the microfauna of the Orinoco-Trinidad-Paria shelves is part of the results of a much wider investigation into the surface sediments of the Orinoco delta and adjoining regions. This research was enabled by the combined efforts of the Institute for Marine Geology of the University of Groningen under the direction of Prof. Dr. PH. H. KUENEN, and the Bataafsche Petroleum Maatschappij N.V. at The Hague with Ir. A. VAN WEELDEN in charge of its Geology Research Department. In addition, several other institutions and people shared in the entailing activities.

The first paper, on the sediment and fauna of the Gulf of Paria, was published by TJ. VAN ANDEL, H. POSTMA and collaborators in 1954. The microfauna of the Gulf was dealt with in this publication by C. KRUIT and A. J. KEY. Further papers on the sediments of the open shelves are those of D. J. G. NOTA (Orinoco-Essequibo) and B. W. KOLDEWIJN (Trinidad-Paria).

The Foraminifera of the latter two regions are reported in the present paper. They were derived from some 150 samples of surface sediment, taken by grab sampler or core sampler during a marine survey from the end of March to early May 1953 (map 1). Only the top part of core samples has been included in the present investigation. The microscopic investigation was carried out in the Paleontological Department of the State University of Utrecht from January 1955 till April 1957.

Object of the investigation was to determine the species, and to examine the possible connection of their occurrence with environmental factors, for the purpose of obtaining a basis for ecological interpretation of fossil associations.

The authors are grateful to Prof. Dr. G. H. R. VON KOENIGSWALD for allowing the facilities of his department. The discussions, especially with Prof. Dr. D. J. DOEGLAS, and Drs. J. R. VAN DER FLIERT, C. KRUIT, A. J. KEY, B. W. KOLDEWIJN and D. J. G. NOTA, are considered of great help.

The Bataafsche Petroleum Maatschappij N.V. (Royal Dutch/Shell Group) gave their kind permission to publish these results. Their ample financial support, facilitating the publication, is highly appreciated.

The drawings of the Foraminifera were made by Mr. F. K. BARENDSE (Utrecht), the maps at the laboratory of the B.P.M. at The Hague.

Representative collections have been stored in the laboratories of the B.P.M. at The Hague, and of the Mineralogisch-Geologisch Instituut of the State University of Utrecht (S 5819-6368; including the figured specimens).

Paleontological Dept. of the State University of Utrecht.

Utrecht, October 1957.

INTRODUCTION

Determination procedure

Over 175 species, subspecies and varieties of Foraminifera were identified. Specific names were first of all looked for in previous descriptions of faunae from the Gulf of Mexico (chiefly those of FRANCES PARKER), in order to facilitate comparison of ecological data. Other names were used only when the earlier identifications were decidedly at variance with our own ideas.

The limits of some of our species are wider than those employed by most American authors, owing to the fact that in several cases of continuous variation we made no specific differentiation. On the other hand, an attempt was made to distinguish variants with differing distribution within the limits of some of the species. Owing to the long duration of the investigation, this endeavour has not always succeeded.

Remarks on determination and distribution, as well as the main references, are given in the systematic description. The species have been arranged alphabetically. Because of technical difficulties, only the more important ones could be figured. For the others references are cited, in which figures can be found that conform to our specimens.

Generic names have occasionally been used in a wide, conservative sense, such as *Rotalia* instead of *Streblus*, etc.

Although no full determinations could be made, a few remarks have also been appended regarding the most commonly occurring species of **Ostracoda.**

Faunal counts

The method used was not based on a constant weight or volume of sediment. The absolute numbers of individuals per species found in this way, are highly dependent on the rate of sedimentation. Consequently, the figures obtained for the various samples, are not comparable.

In practice, much more satisfactory results are obtained by the method of percentage figures, i.e. relative frequencies.

The basis taken for the present investigation was 200 benthonic specimens per sample. Procedure was as follows. The washed residue (particles greater than 60 μ), or a representative part of it, was strewn on a tray without holes. The first 200 benthonic specimens encountered were determined, counted per species, and the results recorded on lists. All planktonic specimens, and also the Ostracoda, found in counting up to 200 benthonic individuals were likewise determined and counted. The endeavour was made to effect an approximately constant, evenly distributed, volume of residue on the tray, for all samples, so that the area covered by 200 benthonic specimens might constitute a rough measure of the relative quantity of sediment (naturally, only as far as the coarser fractions were concerned).

In a number of samples, fewer than 200 benthonic individuals were available. To obtain comparable figures, the frequencies were converted into terms of 200 and the results recorded on the charts in a special way. The deviating numbers of benthonic individuals counted per sample, are shown in a separate list (see explanation of the maps at the end).

Representation of the quantities counted

In this way we have found relative frequencies for every species in the case of each sample. Further procedure differs from that of most American authors (except PHLEGER, 1952) in that the values found were not correlated mainly with depth, but were plotted on a separate map for each species. On the basis of these figures (based on the 200 specimens) contour lines of uniform relative frequency were drawn, generally the lines for 4, 10, 20, 40, 60, etc. These figures divided by two yield percentage figures of the benthos. The figures for planktonic forms and for Ostracoda are proportional figures without this property.

For correct interpretation of the maps, only the more important part of which can be reproduced, the following points should be borne in mind.

1. The figures given are proportional figures, i.e. for each species in relation to the rest of the benthonic fauna. Consequently, they partly depend on the development of other species at the place concerned. In interpreting the maps, the observer might develop a tendency ultimately to see the figures as absolute frequency values. Throughout the rest of the text, the frequencies have not always been qualified by the word "relative", but it must nevertheless constantly be borne in mind that they *are* relative.

2. The individual relative frequencies may vary greatly in real value. In fig. 1, on the assumption of random distribution of individuals of all species on the tray, the connection has been shown between the values found and the real values, for probabilities of $P = 0.045$ and $P = 0.003$ (based on the theory of sampling limits according to formula $N_p \pm (2$ or $3 \sqrt{Npq})$. Therefore, with $P = 0.003$, a found value of 25 % may actually lie between 17 and 35 %. A found frequency of 4, on which many contour lines have been based, can consequently vary from approximately 15 to almost 0, i.e. is of little importance in itself (a frequency of 4 over a larger area has, of course, more value). Accordingly, this evaluation of the individual frequencies must also constantly be borne in mind in carrying out the interpretation. The contour lines have always been drawn as fully as possible. Their real value can therefore differ widely.

3. From the method adopted, it also follows that the relative maxima of various species cannot, as a rule, be expected to coincide, unless we have to do with a definite biotope, greatly differing from those lying around it.

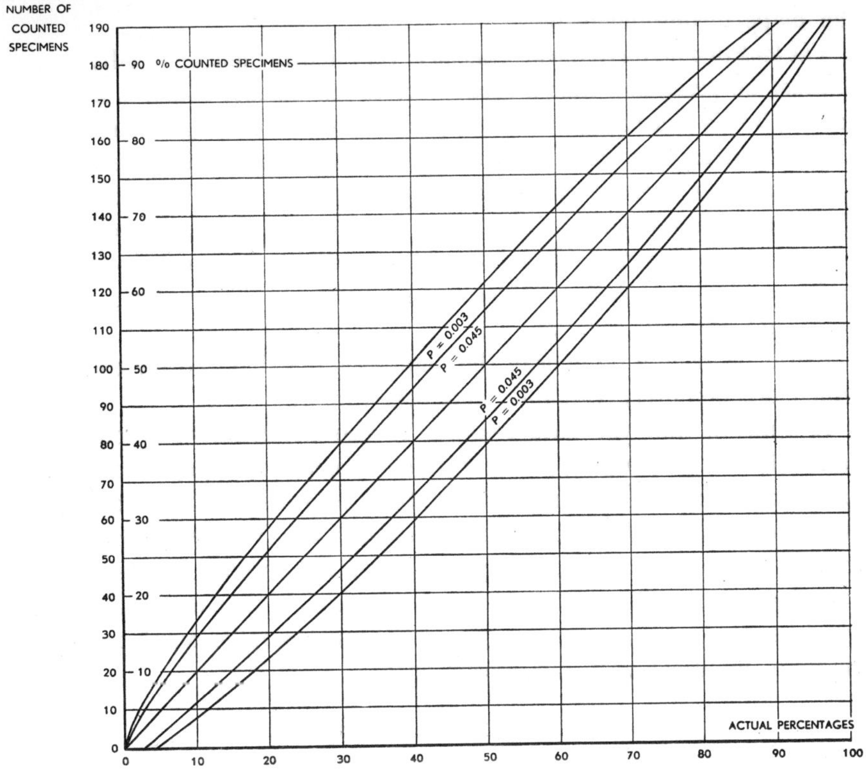

Fig. 1

Relation between observed and real quantities of counted individuals (based on a total of 200), and between the corresponding percentages.

As a consequence the maps are of widely differing value. A low value may be due to various causes.

1. The individuals of the species are not sufficiently numerous.

2. For unknown reasons, the species is very arbitrarily distributed over the area. This is very often the case with the species mentioned under 1, but it also occurs repeatedly with the more numerous ones.

3. Owing to the long duration of the investigation, the boundaries between the different types within one biological species have not always been interpreted in the same way. This is especially true for the Globigerinidae, where even in a shorter time it would have been hardly possible to obtain more reliable counts, in view of the large number of types and the complete transition between most of them.

But even the clearer charts are not particularly easy to interpret. Here again, there are various reasons. Among the external circumstances, only depth and sediment type are well known (KOLDEWIJN and NOTA, our map 2). Knowledge of another important environmental factor, salinity and its fluctuations, is based almost entirely on suppositions. In addition, too great a distance between the samples, especially in the irregular submarine topography of the Trinidad-Paria shelf, makes many details in the maps incomprehensible.

The lack of data on living specimens is very acutely felt, especially because it must be presumed that a great part of the surface fauna is not really recent, but is older and, moreover, probably of varying age. Consequently, all our conclusions on ancient or fossil occurrences had to be based on hypotheses as regards the areas occupied by the corresponding living organisms. This means that there is a great risk of our reasoning in a vicious circle, the more so as the data of earlier authors probably also embrace fossil associations which were reported as recent. Data on living organisms are still scanty.

However, from a number of the maps it is possible to draw conclusions which are definitely valuable, and partly new.

It appears that the data relating to the environment of our species can only be properly applied in the ecological interpretation of fossil associations on the Atlantic side of the Central-American area. In this Atlantic area interpretations of fossil associations from Oligocene onwards are, in our opinion, possible by means of the recent data (see below). Conclusions regarding more remote associations, both in space and in time, must be much more vague, because of the much smaller number of species that are identical with our recent ones.

Rate of sedimentation

Because of the presumed occurrence of non-recent faunae on the surface in the area examined, it is necessary to estimate the relative rates of sedimentation.

Strong sedimentological evidence points to the conclusion that the sand components in the deposits on the outer part of the Orinoco shelf are probably not recent.

This follows from the fact that, at present, the larger rivers evidently supply material which is mainly fine (silt and pelite), the coarser part of which is deposited not far from the mouth. Only the areas in the vicinity of Trinidad and Tobago might exhibit recent sedimentation of coarser, chiefly organogenic, detritus. Consequently, this marks the outer part of the Orinoco shelf and the surroundings of Cumberland Bank as areas of very slight sedimentation, i.e. where older material still lies on the surface or immediately beneath it.

This does not mean that the fauna is entirely fossil or ancient. Part of it will be recent, which conforms, sedimentologically, with the observed admixture of recent pelite (NOTA, 1958). The presence of the recent fauna admixtures appears, for instance, from such features as the fairly gradual increase in the ratio figures for plankton in offshore direction. The degree of admixture undoubtedly varies, but it is very difficult, if not impossible, to ascertain accurately. Differences in conservation are rarely obvious. It is true that, in the outer part of the shelf, we found that a larger proportion of the fauna has been generally more or less strongly corroded, but this is not necessarily a sign of great age.

A number of local faunae are distinctly ancient; this can be proved without trouble. However, as regards the larger areas, there is much more difficulty. The fauna of the outer Orinoco shelf resembles that of the shallower area immediately to the east of Trinidad. The latter is probably recent (provided, we exclude subrecent tectonic elevation of the area). But both regions also resemble one another sedimentologically: sandy with greater or smaller admixture of organogenic detritus (Bryozoa, etc.). Consequently, the coarse sediment, and not the depth, could be the decisive environmental factor for many species. In that case, those species would be occurring in a living state in older sediment.

THE VARIOUS FAUNAL ZONES AND AREAS

It is possible to distinguish a number of distinct zones and areas, each of which has its characteristic species. For those in which recent and older material probably have been mixed, the term area is considered more appropriate than zone.

The Rotalia area along the coast of the Orinoco shelf

Two species, *Rotalia beccarii* and *R. sarmientoi* are strongly predominant off the mouths of the Essequibo and Orinoco, and off the coast in between. This predominance probably results from a much greater tolerance for wide fluctuations of the salt content than is the case with other species. In places where there is little supply of fresh water, such as along the oceanward coasts of Trinidad and Paria, the frequency of these *Rotalia* species appeared to be much smaller.

West of the mouth of the Orinoco, our *Rotalia* strip merges into the near-shore *R. beccarii* zone in the Serpents Mouth (see KRUIT).

A striking feature is that, in offshore direction, the high frequencies of both species adhere to the 20-fathom line, regardless of the sediment type. This outer limit of the distribution can certainly not be explained by equal salinity fluctuations along the entire 20-fathom line. In the case of the Essequibo with its large, open area down to this line, fresh water supply is not greater to that extent than in the area off the Orinoco. A connection with the depth of light penetration would be possible if the species were dependent on, for instance, symbiotic algae, a possibility which is not to be excluded. A connection with coarser, sandy and silty sediment might also exist.

Another explanation for the remarkable distribution, and one to which, in our opinion, preference must be given, is that the surface material of the outer Essequibo "delta" is older. The *Rotalia* association would not be recent either there.

The frequency of *R. sarmientoi* is somewhat higher off the Essequibo than it is off the Orinoco. For a long time we have thought that this indicated that the species does not occur in a living state in this area. This assumption was considered supported by the dull appearance of the test and by the fact that the species, originally described from the Miocene, is not otherwise known to occur living.

However, the relative frequency of *R. sarmientoi* and *R. beccarii* can probably be better explained by *R. sarmientoi* still surviving and having a greater preference than *R. beccarii* for more sandy material. The fairly uniform distribution of *R. sarmientoi* in the Essequibo area is probably due to resedimentation, as the size of the individuals often appeared to be about equal to the grain size of the sediment. Such resedimentation does not necessarily imply that the species is fossil.

Other data point more decisively to a probable subrecent age of the outer part of the Essequibo "delta". *Rotalia rolshauseni* (and the associated *R. pauciloculata*), so frequent in the Gulf of Paria, are much less numerous on the Orinoco shelf. The occurrence of *R. rolshauseni*, outside the *R. beccarii* zone in the Serpents Mouth, reported by KRUIT, continues off the Orinoco, where the species is found towards the outer border of our *Rotalia* band. A lesser tolerance than *R. beccarii* and *R. sarmientoi* for salinity fluctuations may account for the position of these occurrences. Off the Essequibo, a strip of *R. rolshauseni* occurrences is found in the innermost part of the *Rotalia* area, and in a place where the salinity fluctuations may well correspond with those in the *R. rolshauseni* strip off the Orinoco. In this connection it is also remarkable that the distribution of *R. rolshauseni* and *R. pauciloculata* extends for some distance in outward direction along the submarine channel north of the mouth of the Essequibo.

An indication to the same effect is afforded by the fact that, other species with less tolerance of differences in salt content also lie principally outside the main *Rotalia* area off the Orinoco, but enter it distinctly off the Essequibo, though generally with lower frequencies (*Nonionella atlantica*, *Bolivina lowmani*, *B. striatula spinata*, *Quinqueloculina* spp.). Some of these species might exhibit this distribution for other reasons; but in any case dependence on sediment type is not clear.

The above indicates the probability that a considerable part of the fauna of the outer *Rotalia* area off the Essequibo may be regarded as non-recent, of indefinite Quaternary age.

The Nonionella-Uvigerina zone of the Orinoco shelf

This zone immediately adjoins, and overlaps, the border of the *Rotalia* area, lying mainly in the pelite region off the Orinoco. High frequencies of *Nonionella atlantica* and *Uvigerina peregrina peregrina* are characteristic between 35 and 60 metres. Two groups of species may be distinguished, viz. those associated with the pelite and those which are more dependent on depth and/or salinity.

N. atlantica is obviously the least sensitive species in this zone, as regards both type of sediment and fluctuations in salinity. The maximum frequencies are found in the pelite area, at a depth of approximately 30–50 metres, but towards the east the area covered by the species shifts into the sandier *Rotalia* area off the mouth of the Essequibo. We also find the species in the shallower water near Trinidad and Paria, where variations in salinity are probably slight. Depth is evidently the most important factor for the distribution of this species.

U. peregrina peregrina, on the other hand, is bound very closely to the pelite area, but not to depth. *Virgulina pontoni*, and *Cancris sagra* exhibit the same preference for pelite. Viewed from the shore seawards we find that the highest frequencies off the mouth of the Orinoco lie approximately as follows: *V. pontoni* (25–40 m.), *N. atlantica* (30–50 m.) and *C. sagra* (30–60 m.), *U. peregrina peregrina* (40–70 m.).

Bolivina lowmani is a frequent species in the shallow part, and *B. striatula spinata* is frequent throughout the entire area. *B. lowmani* exhibits its high frequencies in silt to pelite under varying salinity conditions. *B. striatula spinata* is less sensitive to depth, and possibly more dependent on the degree of pelite admixture. A striking point is that both species have fairly high frequencies directly off the mouth of the Essequibo, in the *Rotalia* area. They approximately follow the distribution of *N. atlantica.*

Less frequent species are *Rotalia rolshauseni* and *R. pauciloculata*, the distributional trend of which also resembles that of *N. atlantica. Buliminella elegantissima* is yet another species of this zone, but it is very rare. The high frequencies of *R. pauciloculata* and *Nonionella opima* off the Serpents Mouth are conspicuous, but difficult to understand.

The Uvigerina peregrina zone north of Trinidad and Paria

This area of recent pelite sedimentation in a completely marine environment yields a very striking fauna. *Uvigerina peregrina peregrina* is predominant, but *Höglundina elegans, Eponides regularis* and *Uvigerina proboscidea* are also very characteristic species. *H. elegans* adheres fairly closely to the area of distribution of *U. peregrina; E. regularis* is found especially in the deeper part (approx. 110–150 m.), like *U. proboscidea,* which is slightly more restricted to this outer part. *Bolivina subaenariensis mexicana* and *B. barbata* are frequent at only a small number of stations.

Some of the above species are often stated to be characteristic of deeper water, but it appears that their tolerance for depth is fairly great, and the pelite environment decisive. They may represent a group living on organic detritus in the mud, but under fully marine conditions.

Pelite is probably much more fauna-selective than the various sorts and gradations of coarser sediment.

Several modifications of the *Uvigerina* association occur, especially towards the periphery of the zone.

Close to the land, local admixtures of shallow-living species, such as *Nonionella atlantica*, are found. Especially off the Bocas del Dragon, this species occurs at fairly considerable depths, like a number of others, which we have already met with in the Orinoco pelite area; *Virgulina pontoni, Bolivina lowmani, B. striatula spinata, Rotalia rolshauseni.* Other species also occur here, the presence of which is not understood: *Cassidulina neocarinata* var.

The extension of the pelite area north of Paria, especially in the most westerly sampling run, is very remarkable. There are still large quantities of *Uvigerina peregrina peregrina*. The individuals of the fauna are generally very small, and a great part of them give the impression of being sorted. This applies particularly to the planktonic part, which, proportionately speaking, is much too strongly represented. The influence of colder water is likely; this is evidenced by the admixture of numerous Cassidulinidae and a number of *Globigerina bradyi* (a species originally described from the Antarctic). In the benthonic fauna, there are very high frequencies of *Cassidulina subglobosa subcalifornica*; unfortunately no clear distribution picture is afforded by this form throughout the rest of the area examined. The increase in *Bolivina lowmani*, *B. striatula spinata*, *Nonionella atlantica*, *Quinqueloculina* spp. and *Cancris sagra*, and the absence of *Höglundina elegans*, are again reminiscent of the Orinoco pelite area.

The shallow "reef" zone east of Trinidad

Unfortunately, this area has been very scantily sampled. Especially in the sector of fairly coarse sediments, two fauna elements are conspicuous.

One is *Amphistegina lessonii*, with high frequencies near the klintite shallows of Emerald Shoal. Nearer to the shore of Trinidad we find very large numbers of Miliolidae, especially of the bigger forms, such as *Quinqueloculina lamarckiana* and the *bicostata* variety. The latter, near-shore zone probably swings into the Serpents Mouth along the south coast of Trinidad (see KRUIT).

These two associations give a very vague impression of forereef and backreef areas, which, of course, cannot be delimited owing to the absence of a distinct, prominent reef zone.

Other striking components of these two faunae are *Cibicides pseudoungerianus* with its varieties, and *Textularia candeiana* and other arenaceous species. The first are found especially round the *Amphistegina* area; the second are more frequent in that of *Q. lamarckiana*. Other species present will be discussed in considering the outer part of the Orinoco shelf.

The data round about Tobago are too scanty to enable important conclusions to be drawn. Close to the island the fauna resembles the *Amphistegina* association of Emerald Shoal, but the resemblance is not strongly pronounced.

Distinct non-living associations of the Orinoco shelf

These are found especially in the deeper area off the Orinoco mouth, and along the border of the shelf. They are of decidedly different nature.

The most striking is the *Uvigerina* association of DZ 1177, where the *U. peregrina peregrina* individuals (almost 80 %), like the rest of the fauna (including 3 % *Nonionella atlantica*), have been distinctly corroded ("rusty" appearance). The association is more indicative of circumstances in the shallower part of the region north of Trinidad than of those prevailing in the recent pelite area off the Orinoco. The fauna probably represents an ancient period of active pelite sedimentation in a fully marine environment.

Not far away, a second type is found, as admixture in the thin surface layer of DG 1026. This admixture mainly consists of *Rotalia beccarii* and *Nonionella atlantica*. It is not impossible that recent individuals of these species have been transported from their proper environments close to the Orinoco mouth; but this is not very probable, as intermediate links are lacking. We might have to do here with reworking of older deposits with outcrops in the immediate vicinity. Under the thin surface layer is found a fauna with a slight admixture of *Amphistegina lessonii*, such as we constantly find on the entire outer part of the Orinoco shelf.

Likewise pointing to the presence of older material directly at the surface, are a number of samples which are conspicuous on account of a very small content of planktonic individuals, e.g. DZ 1177 and DG 1026, mentioned above, and also DG 1028 and DZ 1173. The last-mentioned samples contain a very high percentage of *A. lessonii* and other stoutly built species, some of which we know to occur in the shallower area east of Trinidad. DW 1156 also belongs distinctly to the last type.

These samples permit little doubt that older associations really occur at the surface. This, therefore, makes it feasible to assume that recent and older material may have been mixed in many other samples in this area.

The outer area of the Orinoco shelf

Throughout this entire area we find an association of fairly coarse sediments with a fauna which consists largely of bigger individuals. It is remarkable that many species are distributed in fairly regular bands running throughout the whole area.

These bands, moreover, often continue to the area with likewise coarse sediments at partly shallower depths, east of Trinidad. On the Orinoco shelf proper they are usually shallowest off the Essequibo, whilst farther west they are shifted by the pelite influence farther outwards and towards the deeper water. In detail, a fair number of irregularities occur, which are partly connected with the coarseness of the sediment.

The difficulty with this area is that it is impossible to make out whether we have to do here with fossil fauna elements which have been "drowned" and not been covered afterwards, or with recent forms which are less sensitive to depth but are associated with the coarse sediment. Perhaps both factors play a part, as for instance, in the case of *Cibicides pseudoungerianus*, which follows the coarser sediments very closely, but the specimens of which greatly vary in size.

The possibility of greater age applies in the first place to those elements which are generally regarded as components of a shallow water fauna of reef environments. These comprise, first and foremost, *Amphistegina lessonii*, but also rarer accompanying forms, such as *Heterostegina antillarum*,

Carpenteria proteiformis, *Liebusella soldanii* and the Peneroplidae. The very high frequencies of *A. lessonii* round Emerald Shoal, east of Trinidad, probably represent recent occurrences which have been more or less displaced. If this is so, the deeper occurrences on the Orinoco shelf would be fossil. This conclusion is supported by the fact that the highest frequencies coincide with very low plankton frequencies. All specimens of *A. lessonii* in this area might therefore be ancient. The more or less patchy character of the frequency distribution of this species, is very likely partly due to the admixture of larger or smaller quantities of recent species. These are probably, in the first place, the smaller species which accompany the pelite admixture, such as *Bolivina striatula spinata*, *Uvigerina peregrina parvula* and *Seabrookia earlandi*. This admixture is especially evident north of the Orinoco pelite area. Sorting may also play a part in the patchy appearance. The coarser faunae lie perhaps on topographically slightly higher points.

Distinctly recent are a number of species, which follow the shelf border fairly closely, such as *Ehrenbergina spinea*, *Cassidulina neocarinata*, *C. laevigata curvata*, *Bolivina subaenariensis mexicana*, *B. goësi*, *U. peregrina peregrina*, *U. proboscidea*, *U. flinti*, *Planulina foveolata*, *Robulus* spp.

There remains the group of species which follow the coarser sediment, and which often "creep upwards" more or less distinctly near Trinidad. This situation is also found, to a certain extent, off the mouth of the Essequibo.

One such example is afforded by the Miliolidae with their bigger forms (chiefly *Quinqueloculina lamarckiana*). This group is very frequent close to the shore of Trinidad, and no doubt the occurrences at somewhat greater depths in this neighbourhood are mainly allochthonous. The occurrences of bigger forms on the Orinoco shelf might again be fossil, as they cannot have been transported from a shallow recent environment there. Generally, the axis of distribution of *Q. lamarckiana* lies at a slightly shallower depth than that of *A. lessonii*. This is in conformity with the relationship of these two species east of Trinidad, and might therefore also point to the ancient character of the species.

Other species with similar distributions and susceptible of two interpretations are *Cibicides pseudoungerianus*, *Reussella atlantica*, *Miliolinella subrotunda*, *Spiroplectammina floridana*, *Textularia candeiana* and *Bigenerina irregularis*.

As far as Tobago, in the more near-shore part of the area, we find a fairly continuous zone of *Elphidium poeyanum*, which is chiefly connected with depth (down to about 85 m.). *Eponides antillarum* and *Hanzawaia concentrica* have a similar distribution. These species are probably mainly recent.

A distribution with two axes is very conspicuous in some species (*Angulogerina occidentalis*, *Reussella atlantica*); it is best developed off the Essequibo. There are no clear indications that the deeper axes are older.

In any case we are left with the general conclusion that fossil associations may have originated from a mixture of widely differing successive faunae. Relative movements of sea level may be one of the causes for such mixtures.

The Cumberland area

This partial region north of Paria resembles in its fauna the outer Orinoco shelf area, but it is different in so far that no species whatsoever is predominant. It is possible that the surface partly consists of older material, but the admixture of recent fauna elements is probably considerable.

INTERPRETATION OF FOSSIL ASSOCIATIONS

An endeavour has been made to interpret ecologically the fossil associations given in three publications on the West Indian region with the help of the data on our recent species. This succeeded rather well in the case of the Miocene and Pleistocene of Aruba, and less well in the case of the other two regions.

Miocene and Pleistocene of Aruba

References: WESTERMANN, J. H. (1951). The water bore of Oranjestad 1942–1943, and its implication as to the geology and geohydrology of the island of Aruba (Netherlands West Indies), I and II. Proc. Kon. Ned. Akad. Wetensch., ser. B, vol. 54, pp. 140–159.

DROOGER, C. W. (1953). Miocene and Pleistocene Foraminifera from Oranjestad, Aruba (Netherlands Antilles). Contr. Cushm. Found. Foram. Res., vol. 4, pp. 116–147.

No sharp division can be made in this borehole by means of our recent associations. This may be due to various causes: impure samples (bailer samples), presence of biotopes not represented in our recent material, erroneous species parallelizations, and differing ecological circumstances affecting species, involved in making the comparison of fossil with recent. Even the possibility that part of the sediment is allochthonous, e.g. as a result of turbidity currents, cannot be entirely excluded.

We shall mainly consider the V, A and C occurrences in DROOGER'S table on pages 134–135.

The lower marly clays of samples 44–39 are superposed on the sterile fine sands of 49–45.

The lowest clay sample, 44, is somewhat differentiated from the others by the presence of great numbers of *Rotalia beccarii* and *Quinqueloculina lamarckiana*. *Uvigerina hispido-costata* (comparable with *U. peregrina peregrina*) is also very numerous. The combination of these three species in our recent area should lie approximately in the overlap zone of the *Rotalia* and *U. peregrina* areas off the Orinoco. A relatively shallow depth would also explain the connection with the sterile fine sands of 49–45 immediately below.

Going upwards, in the series up to and including 39, the great influence of *R. beccarii* and *Q. lamarckiana* disappears, but *U. hispido-costata* remains very frequent. This points to a greater depth (see fig. 2), and corresponds, on our shelf maps, with a shift to the north, on which we must imagine the *Uvigerina* zone of Trinidad connected in outward direction with that of the Orinoco. This assumption would furthermore tally with the ap-

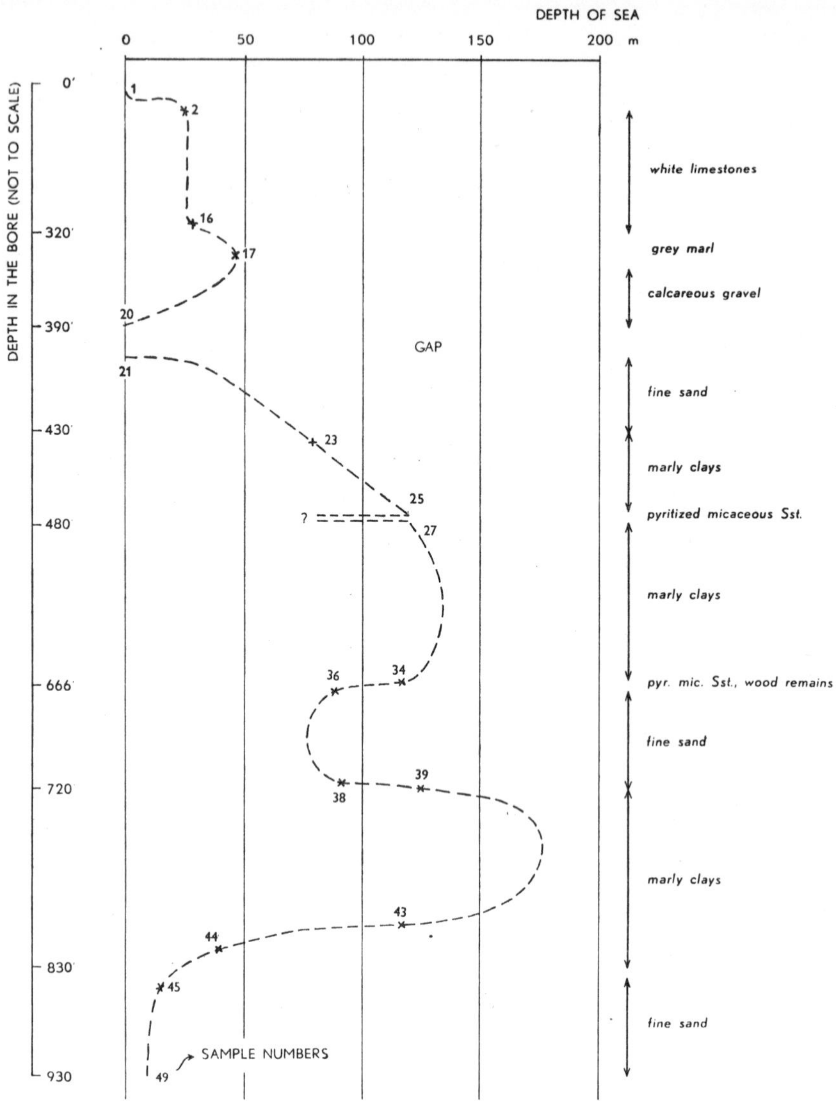

Fig. 2

Estimated sea depth for the successive samples of the Oranjestad borehole, assuming that additional factors (temperature etc.) were similar to those of our recent faunal zones.

pearance of large numbers of *Bolivina subaenariensis westermanni* (cf. *B. subaenariensis mexicana*). The low (but in the bottom samples most regular) frequencies of *Virgulina pontoni*, *Nonion sloani* (cf. *Nonionella atlantica*) and *Bolivina acerosa* (cf. *B. striatula spinata*) indicate the shift through the *Uvigerina* zone off the Orinoco. It conforms with the gradual deepening of the area of sedimentation. Noteworthy is the high frequency of the exclusively fossil *Nonion triangulare* in 44, where one would expect a very great influence of *Nonionella atlantica*. Deeper conditions, comparable with those of the *Uvigerina* zone of Trinidad, might be deducible from the admixtures of *Ehrenbergina spinea*, *Cassidulina carapitana* (cf. *C. laevigata*) and *Gyroidina venezuelana arubana* (cf. *Eponides regularis*), but indistinct forms are also frequent: *Cassidulina subglobosa* with its variety *subcalifornica*, *Elphidium discoidale*, *Bulimina inflata* etc.

In the series of 44 to 39 inclusive, the most acceptable interpretation would appear to be an increase in depth of an area of pelitic sedimentation from about 40 to about 175 m., possibly followed by a substantial decrease in depth.

Samples 38–36, again, are in fine sands. With the change of sediment, *B. subaenariensis* disappears, and *U. hispido-costata* becomes much less frequent. Shallower conditions, such as we find in the seaward-situated marginal area of the shallows of eastern Trinidad, and also on the sandy part of the Orinoco shelf, are conceivable as regards these associations. This would be in conformity with the higher frequencies of *Amphistegina lessonii*, *Cibicides americanus* (cf. *Hanzawaia concentrica*), *Quinqueloculina lamarckiana*, and *Eponides parantillarum* (cf. *E. antillarum*). *Robulus* spp. are less distinct, as are also the high frequencies of *Uvigerina rutila* – evidently a fossil *Uvigerina* with a preference for sandy sediment.

This is followed by the horizon of pyritized sandstone with wood remains of sample 35.

Then follow the upper marly clays of 34–23, with a distinct recurrence of the high *U. hispido-costata* frequencies, but decreasing in an upward direction. The same applies to *Bolivina subaenariensis mexicana*. However, the entire fauna becomes quantitatively poorer, a fact which is connected with the lesser absolute frequencies in the borehole. A conspicuous feature in the deeper samples in this part of the borehole is the admixture of *Nonionella opima*, which might possibly indicate increasing depth from 34 upwards. Otherwise the fauna is not very characteristic, though there are slight differences from that of the lower clays. For instance, although *Epistomina elegans* (= *Höglundina elegans*) is present, none or very few *Ehrenbergina spinea*, *Bolivina acerosa* or *Gyroidina venezuelana arubana* are to be found. This type of admixture might indicate water which was somewhat shallower than in the case of the lower clays.

In the 34–23 interval the fauna becomes impoverished in species and in individuals; this is possibly the result of more rapid sedimentation with an admixture of more sandy material (in 26 pyritized sandstone again), but it may also be a result of increasing shallowness.

The fine sands of 22–21 (23 transitional fauna) are once more of predominantly shallow character, as is evidenced by the large numbers of *A. lessonii*.

After emersion or a period of no sedimentation between 21 and 20, shallow marine conditions recur in the gravel, with a clay admixture higher in the borehole, in samples 20–18. Once again there are large numbers of *A. lessonii*, and also many *Cibicides* and *Planulina*. The frequency of *Epistomina elegans* is probably connected with the clayey character of sample 17.

The top samples (16–2) are from reef limestones, with a poor microfauna containing virtually nothing but *A. lessonii*.

In fig. 2 the estimated sea depth has been recorded for the various samples. The error probably does not exceed 50 % of the value of each point on the curve.

The greater accuracy of this interpretation can be seen from comparison with DROOGER, 1953. On the basis of the data available at that time, a depth of 100–1,000 metres was assumed for the interval covered by the samples 44 to 21.

Oligo-Miocene of the Agua Salada group in Venezuela

Reference: RENZ, H. H. (1948). Stratigraphy and fauna of the Agua Salada Group, State of Falcón, Venezuela. Geol. Soc. Am., Mem. 32.

These faunae differ too strongly from ours for a reasonable comparison to be made. For RENZ' older zones there are too few indications which would permit comparison with the associations of the Orinoco-Trinidad-Paria shelf area. Furthermore, it appears likely that components of several biotopes have been counted together for each stratigraphic unit, while the lower six units have probably been deposited in an environment too deep to permit any comparison at all (200–600 m., according to RENZ).

It is possible that the brackish tinge of the fauna of the top *Elphidium poeyanum-Reusella spinulosa* zonule is slight. This zonule contains *E. poeyanum*, *Rotalia beccarii*, *Reusella spinulosa* (cf. *R. atlantica*), *Cibicides americanus* (cf. *Hanzawaia concentrica*), and *Amphistegina lessonii*. Variations in sea depth of 15 to 40 m. seem to be the most feasible. RENZ quotes depths of less than 30 m., and "subsaline to brackish conditions". In all probability this conclusion is not justified for the zonule as a whole.

More precise information can certainly be obtained from the sample tables given by RENZ.

Miocene Tubara beds of Colombia

Reference: REDMOND, C. D. (1953). Miocene Foraminifera from the Tubara beds of northern Colombia. Jour. Pal. vol. 27, pp. 708–733.

Here again, there is the great disadvantage that the associations of many strata of differing lithology have been combined in the distribution chart.

The lower zone (zone I) was probably deposited in predominantly shallow water to a depth of about 75 metres. The sediments are probably mainly clastic. These conditions may be presumed from the abundant and common frequencies of *Amphistegina lessonii*, *Angulogerina jamaicensis* (cf. *A. occidentalis*), *Buliminella elegantissima*, *Cibicides concentricus* (= *Hanzawaia concentrica*), *C. io* (cf. *C. pseudoungerianus*), *Elphidium* spp., *Nonionella atlantica*, *Quinqueloculina lamarckiana*, and *Rotalia beccarii*, as well as the low frequencies of *Cancris sagra*, *Uvigerina peregrina* and *Virgulina pontoni*. The Globigerinidae are likewise not numerous. This, of course, represents the sum of many strata. The individual depths may have fluctuated between 10 and not much more than 100 m., at any rate if all strata are well represented in the diagram.

Zone II is more or less the same. It probably contains fewer very shallow intercalations, in view of the decline in the frequencies of *A. lessonii*, *B. elegantissima* and *N. atlantica*.

For both zones, REDMOND assumed a neritic environment without further specification.

SYSTEMATIC DESCRIPTION FORAMINIFERA

Benthonic species

Amphistegina lessonii D'ORBIGNY

Map 3

Amphistegina lessonii D'ORBIGNY, 1826, Ann. Sci. Nat., vol. 7, p. 304, pl. 17, fig. 1–4, mod. 98; CUSHMAN, 1931, U.S. Nat. Mus., Bull. 104, pt. 8, p. 79, pl. 16, fig. 1–3; PHLEGER and PARKER, 1951, Geol. Soc. Am., Mem. 46, pt. 2, p. 26, pl. 13, fig. 13, 14, pl. 14, fig. 1.

Remarks. Larger specimens are often fairly thick. They resemble D'ORBIGNY's *A. gibbosa*.

Distribution. In CUSHMAN's monograph of the recent Foraminifera of the Philippines (1921) the following statement is given: "Very common especially in warm, shallow water. It is very frequent in waters less than 30 fathoms in depth, and makes up a considerable proportion of the sand on some of the beaches. . . . usually associated with *Heterostegina* . . . The deeper records, of which there are comparatively few, are usually represented by very few specimens". The same impression is got from the works of BRADY (1884) and others. In the monograph on the Atlantic Foraminifera (CUSHMAN, 1931) 14 "abundant" and "common" occurrences are listed. They range from 0.5–56 fath. (1–102 metres), with an average of 15 fathoms (27 m.) (The average is of course dependent on the available samples).

GOULD and STEWART (1956) recognize an *Amphistegina (lessonii)* zone along the western Florida Coast from 30–58 fath. (55–106 m.). The *Amphistegina* association is found on a (Pleistocene) zone of dead algal rock. The corresponding living Algae are found from 0–30 fathoms. The *Amphistegina* specimens were considered to belong to a living association, but BANDY later informed us (personal communication) that living individuals were extremely rare. He shares our opinion that the majority are not recent at this depth. GOULD and STEWART reported another, fossil, *Amphistegina* occurrence from a depth of 98 fath. (179 m.). Other anomalous *Amphistegina* occurrences have been reported mainly on submarine knobs, by LOWMAN (67 m.) and PHLEGER and PARKER (128 m.).

Two different areas may be distinguished in our material. The first one is east and northeast of Trinidad with a maximal frequency of 60 % at a depth of 22 metres, on a calcarenite and klintite bottom. All surrounding frequencies of more than 10 % are shallower than 70 m. This distribution evidently corresponds to that of the earlier authors: warm, shallow water of a reef area, probably with some displacement after death towards greater depths.

The second area, which is on the Orinoco shelf, is more or less restricted to an elongate zone in between 55 and 180 metres, with some of the maxima on or near the continental slope. It follows the very sandy zone, which is more or less rich in calcarenite. This distribution is equivalent to that recorded by GOULD and STEWART, but there is considerable extension towards greater depths, where all authors would consider the Amphisteginae as not-living.

Since there is on the Orinoco shelf no accompanying shallow zone of the first type, we would prefer to regard all these assemblages as fossil. The same is true of the single occurrence north of the Bocas del Dragon and possibly of part of the deeper samples in the area east of Trinidad.

Angulogerina occidentalis (CUSHMAN)

Map 4

Uvigerina angulosa, CUSHMAN (not WILLIAMSON), 1922, Carnegie Inst. Wash., Publ. 311, p. 34, pl. 5, fig. 3, 4.

Uvigerina occidentalis CUSHMAN, 1923, U.S. Nat. Mus., Bull. 104, pt. 4, p. 169.

Angulogerina occidentalis (CUSHMAN), CUSHMAN, 1930, Florida State Geol. Survey, Bull. no. 4, p. 50, pl. 9, fig. 8, 9; DROOGER, 1953, Contr. Cushm. Found. Foram. Res., vol. 4, p. 136, pl. 21, fig. 32, 33.

Remarks. Our *Angulogerina* specimens are highly variable as to the relative length of the test and to the degree of ornamentation. Usually the greatest width of the test is in the lower part, the later chambers becoming more remote and triangular. Short variants resemble *A. baggi* (GALLOWAY and WISSLER) and *A. hughesi picta* TODD from Californian Pleistocene and recent deposits respectively.

The majority of our specimens resemble the figured individuals from the Miocene of Florida and Aruba better than the recent types from off Tortugas. No doubt *A. bella* PHLEGER and PARKER (1951, Geol. Soc. Am., Mem. 46, pt. 2, p. 12, pl. 6, fig. 7, 8) and *A. jamaicensis* CUSHMAN and TODD, as figured by PARKER (1954, Bull. Mus. Comp. Zool. Harv. Coll., vol. 111, no. 10, p. 521, pl. 8, fig. 8), both from the Gulf of Mexico, are represented among our material, often together in a single sample. Intergradation of these various types made a clear separation into different species impossible.

Distribution. In PHLEGER and PARKER (1951) *A. bella* is said to be most characteristic of depths between 50 and 120 metres. PHLEGER (1956) recorded *A. bella* living at most of his stations between 24 and 100 metres.

East of Trinidad there is a distinct high frequency band between 40 and 60 fath. (73–110 m.). A wider band occurs in the eastern part of the Orinoco shelf, where two high frequency axes appear on the map, one approximately following the 20 fath. line (37 m.), the other that of 40 fath. (73 m.) with occasional deeper occurrences (e.g. 183 m.). The latter band might correspond to the distribution east of Trinidad. There is no apparent relation with sediment type; the species is commonly absent from the pelitic zones.

Articulina sagra d'Orbigny

Articulina sagra d'Orbigny, 1839, in Ramon de la Sagra, Hist. Fis. Pol. Nat. Cuba, "Foraminifères", p. 160, pl. 9, fig. 23–26; Cushman, 1929, U.S. Nat. Mus., Bull. 104, pt. 6, p. 51, pl. 11, fig. 7; Phleger and Parker, 1951, Geol. Soc. Am., Mem. 46, pt. 2, p. 7, pl. 3, fig. 8–10.

Remarks. Only a few specimens.

Astacolus crepidulus (Fichtel and Moll)

Nautilus crepidula Fichtel and Moll, 1803, Test. Micr., p. 107, pl. 19, fig. g–i.

Cristellaria crepidula (Fichtel and Moll), H. B. Brady, 1884, Rep. Voy. Challenger, Zoology, vol. 9, p. 542, pl. 67, fig. 17, 19, 20; Cushman, 1923, U.S. Nat. Mus., Bull. 104, pt. 3, p. 117, pl. 35, fig. 3, 4.

Remarks. Brady's interpretation of this species is sufficiently wide to incorporate our rare individuals.

Distribution. Our scattered occurrences (freq. 1 % or less) are nearly entirely restricted to depths of 95 m. and deeper. There are two occurrences in the DB-traverse at 48 and 53 m.

Bifarina advena (Cushman)

Pl. 1, fig. 1

Siphogenerina advena Cushman, 1922, Carnegie Inst. Wash., Publ. 311, vol. 17, p. 35, pl. 5, fig. 2.

Bifarina advena (Cushman), Key, 1954, Verh. K.N.A.W., ser. 1, vol. 20, no. 5, p. 213, pl. 2, fig. 6.

Bifarina decorata Phleger and Parker, 1951, Geol. Soc. Am., Mem. 46, pt. 2, p. 12, pl. 6, fig. 9, 10.

Remarks. Most of our specimens are rather coarsely ornamented in the biserial portion of the test, thus being closest to the specimens figured by Phleger and Parker.

Distribution. According to Phleger and Parker this species occurs as deep as 205 m., but it is most frequent in water shallower than 100 m. The species was originally described from shallow water of the Tortugas region. In the Gulf of Paria "the species seems to be most prominent in muds of 35 m. on the submarine delta slope" of the Serpents Mouth (Kruit, 1954).

In our material low frequencies, 5 % or less. Scattered maximal occurrences between 40 and 80 metres, commonly about 55 m. There are some isolated occurrences near the continental slope (110 m.). There is no clear dependence on sediment type.

Bigenerina irregularis PHLEGER and PARKER

Map 5

Bigenerina irregularis PHLEGER and PARKER, 1951, Geol. Soc. Am., Mem. 46, pt. 2, p. 4, pl. 1, fig. 16–21; BANDY, 1954, U.S. Geol. Survey, Prof. Paper 254–F, p. 135, pl. 29, fig. 8, 9; PARKER, 1954, Bull. Mus. Comp. Zool. Harv. Coll., vol. 111, no. 10, p. 492, pl. 3, fig. 1–3.

Remarks. When our material from different samples is compared, considerable differences appear in average size of the individuals. The biserial portion of the test is variously compressed.

In one sample (DC 1014 near Bocas del Dragon) small immature specimens predominate. They commonly consist of no more than the textularian stage. In the other samples individuals with several uniserial chambers are much more common.

The differences between *B. irregularis* and *B. textularioidea* (GOËS), reported by PARKER from the Gulf of Mexico (1954, p. 492), are of no use in our adult individuals.

Distribution. The remarks on the distribution of *B. irregularis* in the Gulf of Mexico are fairly homogeneous.

PHLEGER and PARKER (1951): confined to water shallower than 100 m., although in one place at 115 m.; usually less than 5 %, occasionally more than 15 %.

PARKER, PHLEGER and PEIRSON (1953): down to a depth of 117 m.

PARKER (1954): frequencies up to 5 % at less than 100 m., deeper (down to 185 m.) less than 1 %.

BANDY (1954): 0–15 %, fluctuating frequency distribution that is not correlative with any of the environmental factors. Observed area down to 45 m.

PHLEGER (1956): living specimens in abundance at most stations between 7 and 101 m.

In our material there is an elongated narrow zone along the entire Orinoco shelf, beginning north of the Essequibo at approx. 50 m. and gradually shifting to deeper water in western direction, to about 100 m. on the average. There are some deeper occurrences on the continental slope; one is isolated with approx. 10 % north of the Essequibo. There is no relation with sediment type.

Bolivina albatrossi CUSHMAN

Bolivina albatrossi CUSHMAN, 1922, U.S. Nat. Mus., Bull. 104, pt. 3, p. 31, pl. 6, fig. 4; CUSHMAN, 1937, Cushm. Lab. Foram. Res., Spec. Publ. no. 9, p. 153, pl. 18, fig. 22–24; PHLEGER and PARKER, 1951, Geol. Soc. Am., Mem. 46, pt. 2, p. 12, pl. 6, fig. 15, 16; PARKER, 1954, Bull. Mus. Comp. Zool. Harv. Coll., vol. 111, no. 10, p. 513, pl. 7, fig. 13.

Remarks. But a few specimens, which only partly can be called identical with those figured by CUSHMAN.

Distribution. Rare and with scattered occurrences at all depths.

Bolivina barbata Phleger and Parker

Pl. 1, fig. 2

Bolivina barbata Phleger and Parker, 1951, Geol. Soc. Am., Mem. 46, pt. 2, p. 13, pl. 6, fig. 12, 13; Parker, 1954, Bull. Mus. Comp. Zool. Harv. Coll., vol. 111, no. 10, p. 514, pl. 7, fig. 14.

Remarks. There is little doubt that our specimens are identical with those from the Gulf of Mexico. In some samples they merge into our *B. marginata*.

Single specimens in a few samples can be assigned to *B. minima* Phleger and Parker, (1951, Geol. Soc. Am., Mem. 46, pt. 2, p. 14, pl. 6, fig. 22, 25, pl. 7, fig. 1, 2). They clearly show the coarsely perforated areas in the lower half of the chambers.

Distribution. In the Gulf of Mexico *B. barbata* is reported as rare with a very wide depth distribution. Parker (1954) found scattered occurrences, mostly shallower than 200 m., and a maximum frequency of 26 % at 155 metres. In the Gulf of Paria the species occurs in the area of the Bocas del Dragon and the NW Gulf. According to Kruit it is present only in marine sands at depths of 25 m. at least.

In our material also scattered occurrences, mostly in pelite or pelitic sediments. The shallowest occurrence is at 55 m., the deepest well beyond the 100 fath. line (183 m.). Some frequencies up to 5 % are found in the mud area north of the Bocas del Dragon (DC 1013, 1011).

Bolivina goësi Cushman

Pl. 1, fig. 3

Bolivina goësii Cushman, 1922, U.S. Nat. Mus., Bull. 104, pt. 3, p. 34, pl. 6, fig. 5; Cushman, 1937, Cushm. Lab. Foram. Res., Spec. Publ. no. 9, p. 154, pl. 18, fig. 25; Phleger and Parker, 1951, Geol. Soc. Am., Mem. 46, pt. 2, p. 13, pl. 6, fig. 17; Parker, 1954, Bull. Mus. Comp. Zool. Harv. Coll., vol. 111, no. 10, p. 514, pl. 7, fig. 16.

Distribution. In the Gulf of Mexico low frequencies between 110 and 220 m. (Phleger and Parker, 1951), and between 135 and 320 m. with maximum 7 % at 256 metres (Parker, 1954).

In our material the distribution is similar, along the continental slope: 80 to more than 200 m.; low frequencies, maximum 7 % at more than 183 metres.

Bolivina hastata Phleger and Parker

Bolivina hastata Phleger and Parker, 1951, Geol. Soc. Am., Mem. 46, pt. 2, p. 13, pl. 6, fig. 18, 19.

Remarks. At a few stations there are specimens that would be identical with our *B. paula*, if they had not a number of low costae on the early part of the test. These specimens, referred to as *B. hastata*, are probably variants of *B. paula*.

Bolivina lowmani PHLEGER and PARKER

Map 6

Bolivina lowmani PHLEGER and PARKER, 1951, Geol. Soc. Am., Mem. 46, pt. 2, p. 14, pl. 6, fig. 20, 21; PARKER, 1954, Bull. Mus. Comp. Zool. Harv. Coll., vol. 111, no. 10, p. 515, pl. 7, fig. 21.

Remarks. The small specimens of this species are sometimes inseparable from immature individuals of *B. striatula spinata* and from slender variants of *B. paula.*

Distribution. According to PHLEGER and PARKER (1951) and PARKER (1954) *B. lowmani* is abundant to common (usually more than 10 %) in their parts of the Gulf of Mexico, widely distributed at all depths. Along the Texas coast PHLEGER (1956) found living specimens relatively common between 9 and 28 metres, occasionally as deep as 110 metres.

In our material there are low frequency occurrences scattered over the entire region. An area of high percentages (up to 30 %) occurs off the mouth of the Orinoco river (down to 60 m.), with increasing frequencies towards the shallow end of the traverses. A similar, less distinct, area is found at the mouth of the Essequibo river. Furthermore there appear to exist concentrations north of Paria peninsula (up to 8 %, down to 95 m.) and north of the Bocas del Dragon (again 8 %, at 125 m.).

All samples with frequent *B. lowmani* have their pelitic character in common. The influence of salinity fluctuations cannot be excluded.

Bolivina marginata CUSHMAN

Pl. 1, fig. 4

Bolivina marginata CUSHMAN, 1918, U.S. Geol. Survey, Bull. 676, p. 48, pl. 10, fig. 1; CUSHMAN, 1937, Cushm. Lab. Foram. Res., Spec. Publ. no. 9, p. 86, pl. 10, fig. 4–6.

Remarks. Because of lack of material for comparison this specific determination is more or less tentative. Our specimens seem to be different from the Miocene Floridan individuals by the commonly more pointed initial end with occasional development of a small initial spine. They lack the costae of *B. subaenariensis* CUSHMAN (1937, id., p. 155, pl. 18, fig. 26-28) and the chambers are commonly less rapidly increasing in relative height. They also resemble *B. cochei* CUSHMAN and ADAMS from the Late Tertiary of California (CUSHMAN, 1937, id., p. 119, pl. 12, fig. 1, 2). Furthermore *B. cuomoi* BOLTOVSKOY (1954, Foraminiferos del Golfo San Jorge, p. 196, pl. 13, fig. 4-6) is close to our specimens, but differs in having the keel only in the early portion of the test.

In some samples the keel of the specimens has a ragged appearance. These specimens seem to be intermediate between *B. marginata* and *B. barbata.* They resemble *B. pisciformis* GALLOWAY and MORREY from the Miocene of Venezuela and Ecuador (CUSHMAN, 1937, id., p. 92, pl. 11, fig. 20, 21). Evidently our *B. marginata* is most closely related to *B. barbata.*

Distribution. In our material scattered low frequency occurrences. In one sample, off the Bocas del Dragon (DC 1013), 8 % at 125 metres in pelite.

Bolivina paula CUSHMAN and PONTON

Pl. 1, fig. 5

Bolivina paula CUSHMAN and PONTON, 1932, Florida State Geol. Survey, Bull. no. 9, p. 84, pl. 12, fig. 6; PARKER, 1954, Bull. Mus. Comp. Zool. Harv. Coll., vol. 111, no. 10, p. 516, pl. 7, fig. 26.

Remarks. The majority of our individuals agree fairly well with the specimen figured by PARKER from the northeastern Gulf of Mexico.

Variation in our material is considerable, approaching or even including the forms of *B. spathulata* (WILLIAMSON) figured by CUSHMAN (1937, Cushm. Lab. Foram. Res., Spec. Publ. no. 9, p. 162, pl. 15, fig. 20-24) from off the British Isles. The latter seem to be different by the slightly more acute periphery, but the lesser curvature of the sutures of *B. spathulata* is very frequent among our specimens as well. The Gulf of Mexico-Caribbean *B. paula* may be a subspecies of *B. spathulata*, hardly different from the species of the northeastern Atlantic.

Some of our specimens seem to be intermediate between *B. paula* and *B. lowmani*, others between *B. paula* and our *B. marginata*.

Distribution. For the northeastern Gulf of Mexico PARKER reported scattered occurrences at all depths and frequencies less than 1 %.

On the Orinoco-Trinidad shelf we found the same distribution, but there are some high frequency zones, one at approx. 20 fath. (37 m.) at the mouth of the Orinoco river, the others in deep water along the continental slope. Determination difficulties may partly account for these widely separated occurrences.

Bolivina plicatella CUSHMAN

Bolivina plicatella CUSHMAN, 1930, Florida State Geol. Survey, Bull. 4, p. 46, pl. 8, fig. 10; CUSHMAN, 1937, Cushm. Lab. Foram. Res., Spec. Publ. no. 9, p. 89, pl. 11, fig. 3, 4, pl. 16, fig. 21.

Remarks. Among our few specimens, typical ones are rare. Some are too elongate, others have the ridges nearly entirely lacking, approaching the type of *B. compacta* SIDEBOTTOM (see CUSHMAN, 1937, pl. 17, fig. 22–24).

Distribution. Scattered occurrences, insignificant frequencies.

Bolivina striatula CUSHMAN var. **spinata** CUSHMAN

Map 7

Bolivina striatula CUSHMAN var. *spinata* CUSHMAN, 1936, Cushm. Lab. Foram. Res., Spec. Publ. no. 6, p. 59, pl. 8, fig. 9; CUSHMAN, 1937, Cushm. Lab. Foram. Res., Spec. Publ. no. 9, p. 155, pl. 18, fig. 32; PHLEGER and PARKER, 1951, Geol. Soc. Am., Mem. 46, pt. 2, p. 14, pl. 7, fig. 7; KEY, 1954, Verh. K.N.A.W., ser. 1, vol. 20, no. 5, p. 212, pl. 2, fig. 4.

Remarks. Most of our specimens closely resemble the figures of this variety; they clearly show the upper part of the chambers without pores.

Generally they lack a pronounced spine. Some specimens are broader, approaching the type of *B. striatula* (see CUSHMAN, 1937, p. 154, pl. 18, fig. 30, 31). Small specimens of the *B. striatula* group with but slight or no ornamentation seem to grade into our *B. lowmani*.

Especially at stations with but few representatives variation may be fairly wide, including specimens that are very close to *B. fragilis* PHLEGER and PARKER (1951, p. 13, pl. 6, fig. 14, 23, 24). In these specimens the striae are fewer or inconspicuous, the test is more compressed, often slightly keeled and the unperforated areas in the upper part of the chambers are usually lacking (pl. 1, fig. 6). In some samples specimens without striae were found as well. They closely resemble *B. lanceolata* PARKER (1954, Bull. Mus. Comp. Zool. Harv. Coll., vol. 111, no. 10, p. 514, pl. 7, fig. 17–20). In our material these variants that resemble *B. fragilis* and *B. lanceolata* cannot be clearly separated from *B. striatula spinata*. They have been included in the countings.

Distribution. According to PHLEGER and PARKER (1951) the variety has a depth range of 30 to 225 m., usually abundant only to 120 m.; according to PARKER (1954) shallower than 240 m. Recorded frequencies up to 5 % in both papers. PHLEGER (1956) reports living individuals along the Texas coast very abundant between 56 and 75 metres, sparser elsewhere (e.g. up to 15 m.). The records of *B. fragilis* are always very similar: abundant living specimens from 53–99 metres.

In the Gulf of Paria *B. striatula spinata* is reported common to abundant in the area of Bocas del Dragon. KRUIT concluded: marine, but stands a decrease of salinity. The usually small specimens may have been re-deposited later.

Our map shows a complicated pattern. Most characteristic are two areas which contain the highest frequencies. Both are mainly between 20 and 40 fathoms (37–73 m.) in pelite or very pelitic sediment; one off the mouth of the Orinoco river (up to 25 %), extending far eastwards, the other in the DB traverse (up to 15 %) north of Paria peninsula.

Seawards of the maximal Orinoco area, somewhat irregularly distributed lower frequencies are found down to the continental slope. The higher frequencies in this area may again have some relation with relatively somewhat higher pelite contents.

Pelite is probably not the only dominant factor in the distribution of *B. striatula spinata*, because of the relative rarity of the variety in the mud area north of Trinidad. Low frequency areas occur close to the Bocas del Dragon and west of Tobago.

A remarkable area, with low frequencies, however, is found between 10 and 20 metres off the mouth of the Essequibo. This area may be compared with the shallow part of the main Orinoco area, but a corresponding deeper part is entirely wanting. This might be due to the sandy instead of pelitic character of the sediments in these deeper areas.

Summarizing, *B. striatula spinata* seems to have a fairly wide depth distribution on the shelf (if displacement and fossil occurrences can be excluded), which somehow depends on the pelite contents of the bottom. Maximal frequencies are found in pelitic areas between 40 and 75 metres. Slight salinity fluctuations can evidently be supported.

Bolivina subaenariensis CUSHMAN var. **mexicana** CUSHMAN

Map 8

Bolivina subaenariensis CUSHMAN var. *mexicana* CUSHMAN, 1922, U.S. Nat. Mus., Bull. 104, pt. 3, p. 47, pl. 8, fig. 1; CUSHMAN, 1937, Cushm. Lab. Foram. Res., Spec. Publ. no. 9, p. 157, pl. 18, fig. 20; PHLEGER and PARKER, 1951, Geol. Soc. Am., Mem. 46, pt. 2, p. 15, pl. 7, fig. 8–10; PARKER, 1954, Bull. Mus. Comp. Zool. Harv. Coll., vol. 111, no. 10, p. 517, pl. 7, fig. 33.

Remarks. In our material the variety *B. subaenariensis mexicana* is very well distinguishable from all other representatives of the genus.

Distribution. For the Gulf of Mexico a very wide depth range is given (35–3250 m.). Highest frequencies between 110 and 260 m. with maximum 29 % at 192 m. (PHLEGER and PARKER, 1951), and between 100 and 270 m. with maximum 32 % at 155 m. (PARKER, 1954).

In our material a more or less regular distribution along the continental slope. The shallowest occurrence at 75 m., the highest frequency (10 %) at more than 200 metres.

Bolivina subspinescens CUSHMAN

Pl. 1, fig. 7

Bolivina subspinescens CUSHMAN, 1922, U.S. Nat. Mus., Bull. 104, pt. 3, p. 48, pl. 7, fig. 5; CUSHMAN, 1937, Cushm. Lab. Foram. Res., Spec. Publ. no. 9, p. 157, pl. 19, fig. 1–3; PHLEGER and PARKER, 1951, Geol. Soc. Am., Mem. 46, pt. 2, p. 15, pl. 7, fig. 11, 12.

Remarks. At one end of the range of variation in our scarce material, highly ornamented specimens resemble the biserial part of our *Bifarina advena*, from which they differ by the greater size and by never becoming uniserial. On the other side very smooth specimens resemble *Bolivina spinescens* CUSHMAN, as figured from Porto Rico (1937, pl. 18, fig. 19). Possibly the latter specimens represent a distinct species.

Distribution. Scattered occurrences and low frequencies. Shallower than 50 metres only some occurrences of our *B. spinescens*.

Bolivina tortuosa BRADY

Bolivina tortuosa H. B. BRADY, 1884, Rep. Voy. Challenger, Zoology, vol. 9, p. 420, pl. 52, fig. 31, 32; CUSHMAN, 1937, Cushm. Lab. Foram. Res., Spec. Publ. no. 9, p. 133, pl. 17, fig. 11–19.

Distribution. A few isolated occurrences at the continental slope.

Bulimina marginata D'ORBIGNY

Pl. 1, fig. 8

Bulimina marginata D'ORBIGNY, 1826, Ann. Sci. Nat., ser. 1, vol. 7, p. 269, pl. 12, fig. 10–12; CUSHMAN and PARKER, 1947, U.S. Geol. Survey, Prof. Paper 210–D, p. 119, pl. 28, fig. 5, 6; PHLEGER and PARKER, 1951, Geol. Soc. Am., Mem. 46, pt. 2, p. 16, pl. 7, fig. 27, 28; PARKER, 1954, Bull. Mus. Comp. Zool. Harv. Coll., vol. 111, no. 10, p. 510, pl. 6, fig. 20.

Remarks. Some of the specimens approach the type of *B. aculeata* D'ORBIGNY.

Distribution. Various records from the Gulf of Mexico and elsewhere. PHLEGER and PARKER (1951): characteristic down to about 200 m.; PARKER (1951): along the New Hampshire coast greatest frequency at around 100 m.; PARKER (1954): main range from 75 to 320 m., maximum frequency 33 % at 168 m.; PHLEGER (1956): range along Texas coast 15–110 m., a few living specimens 18–42 m.

According to NATLAND (1950) the species prefers a muddy bottom; it is most abundant between 10 and 275 m.

In our material no distinct pattern. Scattered occurrences at all depths. Maximum frequency (8 %) at more than 183 metres. No distinct preference of mud environment; higher frequencies in pelitic sediments.

Bulimina spicata PHLEGER and PARKER

Bulimina spicata PHLEGER and PARKER, 1951, Geol. Soc. Am., Mem. 46, pt. 2, p. 16, pl. 7, fig. 25, 30; PARKER, 1954, Bull. Mus. Comp. Zool. Harv. Coll., vol. 111, no. 10, p. 510, pl. 6, fig. 22, 23.

Remarks. As to the ornamentation some of our specimens seem to be intermediate between *B. spicata* and *B. marginata* and between *B. spicata* and *B. striata mexicana*. This is especially true of young individuals.

Well developed specimens of our *B. spicata* are hardly different from *B. subornata* BRADY (see CUSHMAN and PARKER, 1947, U.S. Geol. Survey, Prof. Paper 210–D, p. 124, pl. 28, fig. 32, 33), a recent species from the western Pacific. The ornamentation of *B. spicata* might be somewhat heavier.

Distribution. Scattered occurrences with low frequencies, mostly below 75 metres. Greatest frequency 7 % at 82 m., in the deepest part of the depression between Trinidad and Tobago (together with 5 % *B. marginata*).

Bulimina striata D'ORBIGNY var. mexicana CUSHMAN

Bulimina inflata SEGUENZA var. *mexicana* CUSHMAN, 1922, U.S. Nat. Mus., Bull. 104, pt. 3, p. 95, pl. 21, fig. 2.

Bulimina striata D'ORBIGNY var. *mexicana* CUSHMAN, CUSHMAN and PARKER, 1947, U.S. Geol. Survey, Prof. Paper 210–D, p. 119, pl. 28, fig. 4; PHLEGER and PARKER, 1951, Geol. Soc. Am., Mem. 46, pt. 2, p. 16, pl. 7, fig. 26, 32; PARKER, 1954, Bull. Mus. Comp. Zool. Harv. Coll., vol. 111, no. 10, p. 511, pl. 6, fig. 24.

Distribution. Shallowest occurrence (out of three) at 135 metres.

Buliminella elegantissima (D'ORBIGNY)

Pl. 1, fig. 9

Bulimina elegantissima D'ORBIGNY, 1839, Voy. Am. mér., vol. 5, pt. 5, Foraminifères, p. 51, pl. 7, fig. 13, 14.

Buliminella elegantissima (D'ORBIGNY), CUSHMAN and PARKER, 1947, U.S. Geol. Survey, Prof. Paper 210–D, p. 67, pl. 17, fig. 10–12; PHLEGER and PARKER, 1951, Geol. Soc. Am., Mem. 46, pt. 2, p. 17, pl. 8, fig. 3, 4.

Distribution. This minute species occurs mainly in shallow water; deeper occurrences are probably all due to displacement. In the Gulf of Mexico optimum frequency above 80 m. (PHLEGER and PARKER, 1951); PHLEGER (1956) found a few living specimens between 7 and 23 m.

In the eastern Pacific most abundant between 18 and 36 m. in Todos Santos Bay, Baja California (WALTON, 1955). Off southern California most abundant between 1 and 25 m., and off the west coast of central America most abundant between 63 and 104 m. (NATLAND, 1950).

In our material some 15 low frequency occurrences, more than half of them shallower than 37 m., but the range extending down to 80 m.

Buliminella parallela CUSHMAN and PARKER

Buliminella parallela CUSHMAN and PARKER, 1947, U.S. Geol. Survey, Prof. Paper 210–D, p. 69, pl. 17, fig. 22.

Distribution. Only four occurrences in the vicinity of Tobago between 50 and 80 metres.

Buliminella semi-nuda (TERQUEM)

Bulimina semi-nuda TERQUEM, 1882, Mém. Soc. Géol. France, ser. 3, vol. 2, p. 117, pl. 12, fig. 21.

Buliminella semi-nuda (TERQUEM), CUSHMAN and PARKER, 1947, U.S. Geol. Survey, Prof. Paper 210–D, p. 61, pl. 16, fig. 4, 9.

Remarks. This Eocene species has been recorded from recent deposits of the Caribbean. Our recent specimens may have fewer chambers in the final whorl than the Eocene type (less than 6–8, approx. 4).

Distribution. A very rare species with a depth range of 55–150 metres.

Cancris sagra (D'ORBIGNY)

Map 9

Rotalina sagra D'ORBIGNY, 1839, in Ramon de la Sagra, Hist. Fis. Nat. Cuba, "Foraminifères", p. 77, pl. 5, f. 13–15.

Cancris sagra (D'ORBIGNY), CUSHMAN and TODD, 1942, Contrib. Cushm. Lab. Foram. Res., vol. 18, p. 74, pl. 18, fig. 1–11, pl. 23, fig. 6; KEY, 1954, Verh. K.N.A.W., ser. 1, vol. 20, no. 5, p. 215, pl. 2, fig. 15.

Remarks. No doubt *C. oblonga*, as recorded from the Gulf of Mexico (PHLEGER and PARKER, 1951; PARKER, 1954) is conspecific with our *C. sagra*. Variation of our specimens distinctly includes the types of both species.

Distribution. The maximal range in the Gulf of Mexico is between 20 and 255 metres. Highest frequencies are found by PHLEGER and PARKER (1951) between 46 and 70 m. (up to 10 %); living specimens by PHLEGER (1956) between 10 and 108 m., very abundant between 48 and 73 metres. According to BANDY (1954) the species is absent from shallow water, but fairly common between 20 and 40 metres. According to KRUIT in the Gulf of Paria mostly below 18 m.; the species may bear a considerable decrease of salinity.

Altogether there seem to be maximal frequencies between approx. 40 and 75 metres.

In our material there is a distinct high frequency area off the mouth of the Orinoco river, extending towards the southeastern coast of Trinidad. Maximal frequencies, up to 15 % between 35 and 70 m. This main area roughly coincides with the shallow mud area off the Orinoco mouth. Furthermore there are scattered low frequency occurrences at the same depth but also deeper. KRUIT's statement that the species may bear some salinity decrease is probably correct.

Carpenteria proteiformis GOËS

Pl. 1, fig. 10

Carpenteria balaniformis GRAY var. *proteiformis* GOËS, 1882, Handl. Kongl. Sv. Vet. Ak., vol. 19, no. 4, p. 94, pl. 6, fig. 208–214, pl. 7, fig. 215–219.

Carpenteria proteiformis GOËS, H. B. BRADY, 1884, Rep. Voy. Challenger, Zoology, vol. 9, p. 679, pl. 97, fig. 8–14; CUSHMAN, 1931, U.S. Nat. Mus., Bull. 104, pt. 8, p. 140, pl. 26, fig. 1.

Remarks. A number of distinct, well developed individuals in two samples: DG 1028 (2 % at 80 m.) and DZ 1173 (3 % at approx. 180 m.). These samples also contain very high percentages of *Amphistegina lessonii.*

Distribution. In the Philippines this adherent species is said to be characteristic of coral reef environment (CUSHMAN, 1921). Unfortunately CUSHMAN's reports only give unattached displaced specimens, thus from deeper water.

The deep occurrences in our material together with abundant *A. lessonii*, point to ancient sediments in these two samples.

Cassidulina laevigata D'ORBIGNY

Pl. 1, fig. 11

Cassidulina laevigata D'ORBIGNY, 1826, Ann. Sci. Nat., ser. 1, vol. 7, p. 282, pl. 15, fig. 4, 5; H. B. BRADY, 1884, Rep. Voy. Challenger, Zoology, vol. 9, p. 428, pl. 54, fig. 1 (not fig. 2, 3); PHLEGER and PARKER, 1951, Geol. Soc. Am., Mem. 46, pt. 2, p. 27, pl. 14, fig. 6; PARKER, 1954, Bull. Mus. Comp. Zool. Harv. Coll., vol. 111, no. 10, p. 536, pl. 11, fig. 2.

Remarks. D'ORBIGNY's original figure is rather poor; generally that of BRADY is taken for reference. Our specimens are not clearly separable from the more frequent variety, *C. laevigata curvata*, from which they differ in the alternating chambers which do not reach so far into the umbilical region, while the sutures are less curved. Many of our specimens have the inward part of the chambers with an abrupt curvature towards the umbo, as is typical of the Miocene *C. carapitana* HEDBERG (1937, Jour. Pal., vol. 11, p. 680, pl. 92, fig. 6).

Distribution. In our material scattered, insignificant frequencies. The few higher frequencies show no depth preference: between 20 fath. (4 %) and over 100 fath. (15 %) (35–190 m.).

Cassidulina laevigata D'ORBIGNY var. **curvata** PHLEGER and PARKER

Map 10

Cassidulina curvata PHLEGER and PARKER, 1951, Geol. Soc. Am., Mem. 46, pt. 2, p. 26, pl. 14, fig. 5; PARKER, 1954, Bull. Mus. Comp. Zool. Harv. Coll., vol. 111, no. 10, p. 535, pl. 11, fig. 1.

Remarks. This is one of the larger forms among the compressed Cassidulinae. It differs from typical *C. laevigata* by the more evenly curved chambers that reach farther into the umbilical area. Difference in perforation, as stated by PHLEGER and PARKER, is not distinct in many of our specimens.

Distribution. In the Gulf of Mexico according to PHLEGER and PARKER (1951) ranging from 40 to 700 metres, generally restricted to depths greater than 90 or 100 metres, with frequencies up to 5 %. According to PARKER (1954) frequencies of 1–5 % between 60 and 600 m.

In our material this variety is mainly found on the outer part of the shelf. In the Orinoco-Essequibo area highest frequencies (up to 12 %) are found in the deepest samples, 180 m. and deeper. North of Trinidad and Paria the high frequency area shows several anomalies; it reaches upwards to approximately 80 metres; the relations between frequency and depth are not clear in this area.

Cassidulina neocarinata THALMANN

Pl. 1, fig. 12

Cassidulina laevigata, H. B. BRADY (not D'ORBIGNY), 1884, Rep. Voy. Challenger, Zoology, vol. 9, pl. 54, fig. 2.

Cassidulina laevigata D'ORBIGNY var. *carinata* CUSHMAN (not SILVESTRI), 1922, U.S. Nat. Mus., Bull. 104, pt. 3, p. 124, pl. 25, fig. 6, 7; PHLEGER and PARKER, 1951, Geol. Soc. Am., Mem. 46, pt. 2, p. 27, pl. 14, fig. 7.

Cassidulina neocarinata THALMANN, 1950, Contr. Cushm. Found. Foram. Res., vol. 1, p. 44; PARKER, 1954, Bull. Mus. Comp. Zool. Harv. Coll., vol. 111, no. 10, p. 536, pl. 11, fig. 3.

Remarks. It is considered doubtful whether *C. neocarinata* is specifically different from *C. cushmani* R. E. and K. C. STEWART (1930, Jour. Pal., vol. 4, p. 71, pl. 9, fig. 5) from the Californian Pliocene.

Distribution. In the Gulf of Mexico according to PHLEGER and PARKER (1951): range 40–1000 m., greatest frequencies (up to 22 %) between 100 and 700 metres; according to PARKER (1954): frequencies 1–11 % between 120 and 600 metres.

In our material, with very few exceptions, restricted to water deeper than 90 m., roughly following the continental slope. Remarkable high frequency of 40 % in EC 1199 (177 m.). Another high frequency of 10 % in DR 1104 (hard clay, according to KOLDEWIJN, personal communication).

Cassidulina neocarinata THALMANN var.

Pl. 1, fig. 13

A large number of very small *Cassidulina* specimens resemble those of *C. neocarinata* in number and arrangement of the chambers. They are different by the smaller size, lack of the carina, and the less elongate aperture which is moreover not confined to the inner margin of the final chamber. In some specimens the early chambers seem to be arranged as in *C. norcrossi*, but the later chambers show the arrangement of *C. neocarinata*. Usually there is a fairly wide clear area in the centre of the test.

Intergradation of *C. neocarinata* and this variety is not apparent, which, however, may be due to the striking difference in size of the majority of the individuals.

Distribution. Mainly scattered, low frequency occurrences. Up to 26 % in some samples near the Bocas del Dragon (DC 1013, 1011).

Cassidulina norcrossi CUSHMAN subsp. **australis** PHLEGER and PARKER

Pl. 1, fig. 14

Cassidulina norcrossi australis PHLEGER and PARKER, 1951, Geol. Soc. Am., Mem. 46, pt. 2, p. 27, pl. 14, fig. 8–10.

Distribution. Scattered occurrences, mainly along the continental slope and in the deeper area north of Trinidad and Paria, commonly deeper than 100 metres. Some shallow occurrences in the DB section may be due to colder water north of Paria (The species *C. norcrossi* was described from shallow water off the Greenland coast).

Cassidulina palmerae BERMUDEZ and ACOSTA

Pl. 1, fig. 15

Cassidulina palmerae BERMUDEZ and ACOSTA, 1940, Mem. Soc. Cub. Hist. Nat., vol. 14, p. 57, pl. 9, fig. 6–8.

Distribution. Occurring in many samples in the area west of Tobago, depth range 70–150 metres, frequencies up to 4 %.

Cassidulina subglobosa BRADY

Map 11; pl. 1, fig. 16

Cassidulina subglobosa H. B. BRADY, 1884, Rep. Voy. Challenger, Zoology, vol. 9, p. 430, pl. 54, fig. 17; PHLEGER and PARKER, 1951, Geol. Soc. Am., Mem. 46, pt. 2, p. 27, pl. 14, fig. 11, 12; PARKER, 1954, Bull. Mus. Comp. Zool. Harv. Coll., vol. 111, no. 10, p. 536, pl. 11, fig. 4, 5, (6–9).

Remarks. This species shows enormous variation in our material. Delimiting of boundaries between groups was impossible. Variation especially concerns size, degree of lateral compression and the shape of the aperture. In our material two different types were arbitrarily separated, a more globular and a more compressed one. Typical globular specimens *(C. subglobosa)* occur among individuals of all sizes, though most commonly among the bigger ones. Some very big specimens (pl. 1, fig. 16) show the opaque wall of the individuals in figures 11 and 12 of PHLEGER and PARKER. Smaller specimens are generally more compressed, belonging to the variety *C. subglobosa subcalifornica.* Both types often occur together, sometimes accompanied by distinct intermediate forms. In other samples, however, one type occurs in abundance with exclusion of the other.

Distribution. PHLEGER and PARKER (1951): in some abundance at all depths sampled. PARKER (1954): very abundant, widely distributed, deeper than 80 m. frequencies usually between 5 and 20 %. PHLEGER (1956): living specimens deeper than about 50 metres.

In our material the species is most frequent on the outer part of the shelf, around Tobago and north of Paria, avoiding the muddy platform of Trinidad. The higher frequencies are generally deeper than 75 m., but there are several exceptions. At several places in the deepest parts of the traverses the frequencies are again less, but this is not a consistent feature. Highest frequency 20 % at more than 183 m. north of Bocas del Dragon.

Cassidulina subglobosa BRADY var. **subcalifornica** DROOGER

Map 12; pl. 1, fig. 17

Cassidulina subglobosa BRADY var. *subcalifornica* DROOGER, 1953, Contr. Cushm. Found. Foram. Res., vol. 4, p. 140, pl. 22, fig. 8, 9.

Remarks. In the Gulf of Mexico these more compressed forms were noted by PARKER (1954, pl. 11, fig. 6, 7?, 8?) and probably by PHLEGER and PARKER (1951, p. 27, pl. 14, fig. 13) as *C.* cf. *subglobosa.*

In our material they were found to be identical with *C. subglobosa* var. *subcalifornica,* originally described from the Miocene of Aruba, in which Miocene material the same relation to *C. subglobosa* occurs, as it is found among our recent individuals. Very typical specimens of the variety were found at stations where this form is dominant.

Some very small specimens have the elongate basal aperture of *C. crassa* D'ORBIGNY, as figured by BRADY (1884, pl. 54, fig. 5). They may be specifically distinct.

Distribution. The distribution pattern of this variety is fairly irregular. An irregular, interrupted band (up to 25 %) follows the continental slope of the Orinoco shelf. Another high frequency zone (up to 15 %) is found from 35 to 75 metres, more or less between the mouths of Orinoco and Essequibo. North of Paria in the DB section very high frequencies are found (up to 40 %) in the "cold water" area southeast of the Cumberland Bank. Still other occurrences of some importance further complicate the distribution pattern.

Cibicides lobatulus (WALKER and JACOB)

Nautilus lobatulus WALKER and JACOB, 1798, Adam's Essays, Kanmacher's Ed., p. 642, pl. 14, fig. 36.

Truncatulina lobatula (WALKER and JACOB), H. B. BRADY, 1884, Rep. Voy. Challenger, Zoology, vol. 9, p. 660, pl. 92, fig. 10, pl. 93, fig. 1, 4, 5, pl. 115, fig. 4, 5.

Cibicides lobatulus (WALKER and JACOB), CUSHMAN, 1931, U.S. Nat. Mus., Bull. 104, pt. 8, p. 118, pl. 21, fig. 3.

Remarks. Distinct specimens of *C. lobatulus* are rare; they are usually small. Irregular variants of the *C. pseudoungerianus* group may have been included as well.

Distribution. Scattered occurrences with low frequencies.

Cibicides pseudoungerianus (CUSHMAN)

Map 13; pl. 2, fig. 1

Truncatulina ungeriana, H. B. BRADY (not D'ORBIGNY), 1884, Rep. Voy. Challenger, Zoology, vol. 9, p. 664, pl. 94, fig. 9.

Truncatulina pseudoungeriana CUSHMAN, 1922, U.S. Geol. Survey, Prof. Paper no. 129–E, p. 97, pl. 20, fig. 9.

Cibicides pseudoungerianus (CUSHMAN), CUSHMAN, 1931, U.S. Nat. Mus., Bull. 104, pt. 8, p. 123, pl. 22, fig. 3–7.

Remarks. Especially the types of CUSHMAN's figures 3, 4 and 6 (1931) are abundantly present in our recent material.

In the course of counting we failed to make a clear separation between the numerous *Cibicides* types of specific rank that have been recognized by CUSHMAN, PARKER and others in the Gulf of Mexico-Caribbean region. The artificial clusters around these types appear highly transgressive, especially if the wide size range of all individuals is taken into account. Some larger types, such as *C. corpulentus* PHLEGER and PARKER, and *C. pseudoungerianus* var. *antilleana* DROOGER (= *C. protuberans* PARKER) are undoubtedly present, but distinct younger specimens of these types seem to be absent. The majority of the middle-sized specimens show a similar range of variation as *C. pseudoungerianus*, as figured by CUSHMAN from the Atlantic (1931). The variation in our material also includes such species as *C. deprimus*, *C. mollis*, *C. umbonatus* and *Cibicides* sp. 2, all described by PHLEGER and PARKER from the Gulf of Mexico. Also their *C.* aff. *floridanus* (CUSHMAN) must be incorporated in this group. The minute *Cibicides* sp. 1 PHLEGER and PARKER was occasionally observed. Furthermore distinct specimens of *C. io* CUSHMAN and *C. robertsonianus* (BRADY) are present.

Every one of these types occasionally occurs more or less distinctly in single samples, but especially when additional samples are taken into account, intergradation makes the boundaries between the types illusionary. The entire group of individuals is therefore placed under *C. pseudoungerianus*, although references will be given to the main other types (*C. io*, *C. robertsonianus*, *C. pseudoungerianus* var. *antilleana*).

Distribution. Counting together the published records of the various species, seems to be of little use.

According to NATLAND (1950), *C. pseudoungerianus* is often found attached to seaweeds, which may extend its range. NATLAND found common to abundant occurrences between 45 and 300 metres along the western coast of central and north America.

In our material there are high frequencies (up to 36 %) in a distinct zone on the outer part of the Orinoco shelf, deeper than 55 metres. This high frequency zone further westwards approaches the eastern coast of Trinidad and swings around Tobago. North of Trinidad and Paria high frequencies are again in the offshore part, but there are relatively low frequencies in the mud area north of Trinidad. The highest frequencies of nearly 50 % are found in the area east of Trinidad.

The distribution of the *C. pseudoungerianus* group seems to follow the area of coarser sediments with highest frequencies in the outer part of the shelf. The absence in pelite areas may be due to a lack of sufficiently large attachment surfaces.

The records of the other three variants follow the zone of high frequencies of the entire group.

Cibicides io CUSHMAN

Pl. 2, fig. 4

Cibicides pseudoungerianus (CUSHMAN) var. *io* CUSHMAN, 1931, U.S. Nat. Mus., Bull. 104, pt. 8, p. 125, pl. 23, fig. 1, 2.

Cibicides io CUSHMAN, PHLEGER and PARKER, 1951, Geol. Soc. Am., Mem. 46, pt. 2, p. 30, pl. 16, fig. 5, 6.

Remarks. Typical forms are characterized by the rounded periphery, mostly slightly evolute dorsal side and the channel along the spiral suture.

Cibicides pseudoungerianus (CUSHMAN) var. **antilleana** DROOGER

Pl. 2, fig. 2

Cibicides pseudoungerianus (CUSHMAN) var. *antilleana* DROOGER, 1953, Contr. Cushm. Found. Foram. Res., vol. 4, p. 146, pl. 23, fig. 4, 5.

Cibicides protuberans PARKER, 1954, Bull. Mus. Comp. Zool. Harv. Coll., vol. 111, no. 10, p. 542, pl. 12, fig. 13, 14, 16.

Remarks. This variety with its large, spreading, somewhat irregular later chambers and its large ventral umbilical knob of clear shell substance, is distinctly present in our material.

Cibicides robertsonianus (BRADY)

Pl. 2, fig. 3

Truncatulina robertsoniana H. B. BRADY, 1884, Rep. Voy. Challenger, Zoology, vol. 9, p. 664, pl. 95, fig. 4.
Cibicides robertsonianus (BRADY), CUSHMAN, 1931, U.S. Nat. Mus., Bull. 104, pt. 8, p. 121, pl. 23, fig. 6; PHLEGER and PARKER, 1951, Geol. Soc. Am., Mem. 46, pt. 2, p. 31, p. 16, fig. 10–13.

Remarks. Typical *C. robertsonianus* differs from *C. pseudoungerianus* s. str. in being more evolute both dorsally and ventrally. The evolute appearance of the ventral side is variable; some specimens were found with an involute ventral side (see fig. 3). Some specimens resemble *C. bradyi* (TRAUTH) (*Truncatulina bradyi* TRAUTH, 1918, K. Ak. Wiss. Wien, Math. Naturw. Kl., vol. 95, p. 235 = *T. dutemplei* H. B. BRADY (not D'ORBIGNY), 1884, p. 665, pl. 95, fig. 5). Possibly they represent young individuals of *C. robertsonianus.*

Dentalina communis (D'ORBIGNY)

Nodosaria (Dentalina) communis D'ORBIGNY, 1826, Ann. Sci. Nat., vol. 7, p. 254, no. 35.
Nodosaria communis D'ORBIGNY, H. B. BRADY, 1884, Rep. Voy. Challenger, Zoology, vol. 9, p. 504, pl. 62, fig. 19–22; CUSHMAN, 1923, U.S. Nat. Mus., Bull. 104, pt. 4, p. 75, pl. 12, fig. 3, 4, 15–17.

Remarks. In this species have been included most of our rare *Dentalina* specimens. Some resemble other recent species of this genus.

Distribution (including *D. consobrina*). Scattered occurrences, 50 m. and deeper, low frequencies.

Dentalina consobrina D'ORBIGNY

Dentalina consobrina D'ORBIGNY, 1846, Foram. Foss. Bass. Tert. Vienne, p. 46, pl. 2, fig. 1–3.
Nodosaria consobrina (D'ORBIGNY), H. B. BRADY, 1884, Rep. Voy. Challenger, Zoology, vol. 9, p. 501, pl. 62, fig. 23, 24.

Remarks. Only a few specimens, which are mostly identical with the individuals figured by BRADY as *N. consobrina* var. *emaciata* REUSS (1884, pl. 62, fig. 25, 26).

The genus *Discorbis* is used in a conservative, wide sense.

Discorbis bulbosa PARKER

Pl. 2, fig. 5

"*Discorbis*" *bulbosa* PARKER, 1954, Bull. Mus. Comp. Zool. Harv. Coll., vol. 111, no. 10, p. 523, pl. 8, fig. 10–12.

Distribution. According to PARKER shallower than 100 m. the frequencies may amount to 2 %, deeper (down to 205 m.) frequencies are less than 1 %. PHLEGER (1956) found living specimens between 22 and 49 m.

In our material this species was found at all depths; it is rare west of

Tobago. Frequencies vary up to 5 %; those of 2 % and more are mostly in water shallower than 70 metres.

Discorbis floridana Cushman

Pl. 2, fig. 6

Discorbis floridana Cushman, 1922, Carnegie Inst. Wash., Publ. no. 311, p. 39, pl. 5, fig. 11, 12; Phleger and Parker, 1951, Geol. Soc. Am., Mem. 46, pt. 2, p. 20, pl. 10, fig. 4.

Rosalina floridana (Cushman), Parker, 1954, Bull. Mus. Comp. Zool. Harv. Coll., vol. 111, no. 10, p. 524, pl. 8, fig. 19, 20.

Remarks. The wide range of variation in our material is mainly connected with the considerable difference in size of the individuals. Small specimens usually have less inflated chambers than adult forms. The number of chambers is usually 6 or 7, occasionally 8 to a whorl, instead of 4 to 6. The sutures are more limbate and the umbo usually contains a distinct clear knob. In the countings these smaller individuals were referred to as *D. clara* Cushman (1949, Mém. Inst. Roy. Sci. Nat. Belgique, no. 111, p. 45, pl. 8, fig. 8), because of their resemblance with this recent species which lives off the Belgian coast. In adult specimens the chambers are more inflated, more overlapping dorsally and the later umbilicus is usually open. Since gradation between both forms was apparent with increasing size, *D. clara* of our material is considered to be no more than a variant of *D. floridana.*

Distribution. From the Gulf of Mexico generally recorded between 10 and 60 m.

In our material there is a very irregular distribution pattern, that covers all depths. High frequencies (up to 12 %) show no uniform distribution.

Discorbis floridensis Cushman

Discorbis bertheloti (d'Orbigny) var. *floridensis* Cushman, 1931, U.S. Nat. Mus., Bull. 104, pt. 8, p. 17, pl. 3, fig. 3–5.

Discorbis floridensis Cushman, Phleger and Parker, 1951, Geol. Soc. Am., Mem. 46, pt. 2, p. 20, pl. 10, fig. 5–7.

Rosalina floridensis (Cushman), Parker, 1954, Bull. Mus. Comp. Zool. Harv. Coll., vol. 111, no. 10, p. 525, pl. 8, fig. 28, 29.

Remarks. Especially the adult specimens are completely within the range of variation indicated by the figures of Phleger and Parker; younger specimens are recognizable by the coarse dorsal pores. They somewhat resemble *Hanzawaia concentrica.*

Distribution. Found at all depths, mostly between 40 and 150 m.; frequencies maximally 2 %.

Discorbis cf. **nitida** (Williamson)

Discorbis cf. *nitida* (Williamson), Phleger and Parker, 1951, Geol. Soc. Am., Mem. 46, pt. 2, p. 20, pl. 10, fig. 8.

Remarks. A number of more or less low-conical specimens seem to be identical with the specimen figured by PHLEGER and PARKER from the Gulf of Mexico.

Distribution. Scattered occurrences between 70 and 100 metres, except for one at 28 m. Altogether 9 recordings.

Discorbis parkerae NATLAND

Discorbis parkeri NATLAND, 1950, Geol. Soc. Am., Mem. 43, pt. 4, p. 27, pl. 6, fig. 11.

Rosalina parkerae (NATLAND), PARKER, 1954, Bull. Mus. Comp. Zool. Harv. Coll., vol. 111, no. 10, p. 525, pl. 8, fig. 24, 25.

Remarks. A few scattered specimens seem to belong to this species as figured by PARKER.

Ehrenbergina spinea CUSHMAN

Pl. 2, fig. 7

Ehrenbergina spinea CUSHMAN, 1935, Smiths. Misc. Coll. Wash., vol. 91, no. 21, p. 8, pl. 3, fig. 10, 11; PHLEGER and PARKER, 1951, Geol. Soc. Am., Mem. 46, pt. 2, p. 27, pl. 14, fig. 18; PARKER, 1954, Bull. Mus. Comp. Zool. Harv. Coll., vol. 111, no. 10, p. 537, pl. 11, fig. 12.

Distribution. PARKER (1954) noted four occurrences between 135 and 255 m., and two at 914 and 950 m.

In our material *E. spinea* is mainly confined to the shelf border deeper than 135 metres with maximal frequency of 15 %. Some rare occurrences were found on the inner shelf as shallow as 24 metres.

Elphidium matagordanum (KORNFELD)

Nonion depressula (WALKER and JACOB) var. *matagordana* KORNFELD, 1931, Contr. Dept. Geol. Stanford. Univ., vol. 1, p. 87, pl. 13, fig. 2; PHLEGER and PARKER, 1951, Geol. Soc. Am., Mem. 46, pt. 2, p. 11, pl. 5, fig. 17.

Elphidium matagordanum (KORNFELD), PARKER, PHLEGER and PEIRSON, 1953, Cushm. Found. Foram. Res., Spec. Publ. no. 2, p. 8, pl. 3, fig. 24, 25.

Distribution. In our material scattered occurrences in the middle part of the shelf between 25 and 160 metres.

Elphidium poeyanum (D'ORBIGNY)

Map 14

Polystomella poeyana D'ORBIGNY, 1839, in Ramon de la Sagra, Hist. Fis. Pol. Nat. Cuba, Foraminifères, p. 55, pl. 6, fig. 25, 26.

Elphidium poeyanum (D'ORBIGNY), CUSHMAN, 1939, U.S. Geol. Survey, Prof. Paper 191, p. 54, pl. 14, fig. 25, 26; BANDY, 1954, U.S. Geol. Survey, Prof. Paper 254–F, p. 136, pl. 30, fig. 6.

Remarks. Although considerable variation exists among the representatives of *Elphidium* in our samples, there is probably no more than a single species. Larger specimens with a test without secondary thickening closely resemble D'ORBIGNY's West Indian species *E. poeyanum.* The later chambers are usually somewhat inflated and the periphery is correspondingly lobulate; the umbilicus is small. Especially among smaller specimens there is a tendency towards the shape of typical *E. discoidale* (D'ORBIGNY), another recent species of the West Indies. Just as our *E. poeyanum* they have some 10 chambers in the final convolution. Possibly other *Elphidium* species have been occasionally incorporated, but the numbers of their specimens were too small for a correct determination.

Distribution. According to CUSHMAN *E. poeyanum* is a shallow water species in the West Indies. According to PARKER (1954) shallower than 145 m. PHLEGER (1956) found occasional living specimens between 7 and 44 metres.

In our material there is a fairly constant zone with high frequencies down to a depth of approx. 85 metres. The shallow limit seems to be related with the sediment type; the species evidently avoids the pelitic area off the mouth of the Orinoco. Maximal frequencies (up to 12 %) are mostly between 30 and 50 metres. West of Tobago the species appears to be very scarce.

Entosolenia fasciata (EGGER)

Oolina fasciata EGGER, 1857, Neues Jahrb. Min. Geogn. Geol., p. 270, pl. 5, fig. 12–15.

Lagena quadricostulata, H. B. BRADY (not REUSS), 1884, Rep. Voy. Challenger, Zoology, vol. 9, p. 486, pl. 59, fig. 15.

Remarks. Scattered single specimens. Various types have been included.

Entosolenia marginata (WALKER and BOYS)

Serpula marginata WALKER and BOYS, 1784, Test. Min., p. 2, pl. 1, fig. 7.

Lagena marginata (WALKER and BOYS), H. B. BRADY, 1884, Rep. Voy. Challenger, Zoology, vol. 9, p. 476, pl. 59, fig. 21, 22; CUSHMAN, 1923, U.S. Nat. Mus., Bull. 104, pt. 4, p. 35, pl. 6, fig. 9.

Remarks. Single-keeled specimens of various types have been included in this species.

Distribution. Low frequency occurrences, all west of Tobago.

Entosolenia orbignyana (SEGUENZA)

Fissurina orbignyana SEGUENZA, 1862, Dei ter. Terz. del distr. Messina, pt. 2, p. 66, pl. 2, fig. 25, 26.

Lagena orbignyana (SEGUENZA), H. B. BRADY, 1884, Rep. Voy. Challenger, Zoology, vol. 9, p. 484, pl. 59, fig. 25, 26; CUSHMAN, 1923, U.S. Nat. Mus., Bull. 104, pt. 4, p. 39.

Remarks. Generally fairly large individuals. The majority of them agree best with the variety *elliptica* (CUSHMAN) (1923, U.S. Nat. Mus., Bull. 104, pt. 4, p. 42, pl. 6, fig. 10–12).

Distribution. The larger individuals are confined to water shallower than 80 metres.

Entosolenia orbignyana (SEGUENZA) var. **lacunata** (BURROWS and HOLLAND)

Lagena castrensis, H. B. BRADY (not SCHWAGER), 1884, Rep. Voy. Challenger, Zoology, vol. 9, p. 485, pl. 60, fig. 1, 2.

Lagena lacunata BURROWS and HOLLAND, in Jones, 1895, Pal. Soc., p. 205, pl. 7, fig. 12.

Lagena orbignyana (SEGUENZA) var. *lacunata* (BURROWS and HOLLAND), CUSHMAN, 1913, U. S. Nat. Mus., Bull. 71, pt. 3, p. 43, pl. 20, fig. 1.

Distribution. The low frequency occurrences of this variety are restricted to depths of 80 metres and shallower.

Epistominella decorata (PHLEGER and PARKER)

Pseudoparrella decorata PHLEGER and PARKER, 1951, Geol. Soc. Am., Mem. 46, pt. 2, p. 28, pl. 15, fig. 4, 5.

Epistominella decorata (PHLEGER and PARKER), PARKER, 1954, Bull. Mus. Comp. Zool. Harv. Coll., vol. 111, no. 10, p. 533, pl. 10, fig. 18, 19.

Remarks. Our specimens assigned to this species, are highly variable, the periphery being rounded to acute. The ornamentation is mainly along the sutures and the periphery.

Distribution. A few occurrences at all depths, frequencies 1 % or less.

Epistominella cf. **E. vitrea** PARKER

Cf. *Epistominella vitrea* PARKER, in PARKER, PHLEGER and PEIRSON, 1953, Cushm. Found. Foram. Res., Spec. Publ. no. 2, p. 9, pl. 4, fig. 34–36, 40, 41; PARKER, 1954, Bull. Mus. Comp. Zool. Harv. Coll. vol. 111, no. 10, p. 534, pl. 10, fig. 20, 26.

Remarks. Our small specimens seem to be more globular than PARKER's individuals.

Distribution. Very few, low frequency occurrences at 40 m. and deeper.

Eponides antillarum (D'ORBIGNY)

Map 15

Rotalina antillarum D'ORBIGNY, 1839, in Ramon de la Sagra, Hist. Fis. Pol. Nat. Cuba, Foraminifères, p. 75, pl. 5, fig. 4–6.

Eponides antillarum (D'ORBIGNY), PHLEGER and PARKER, 1951, Geol. Soc. Am., Mem. 46, pt. 2, p. 20, pl. 10, fig. 9, 10; PARKER, 1954, Bull. Mus. Comp. Zool. Harv. Coll., vol. 111, no. 10, p. 528, pl. 9, fig. 14, 15.

Remarks. The young specimens of this species are rather different from adult ones, which are mostly thick with numerous chambers. The juveniles are usually more flattened dorsally. Some variants resemble *E. schreibersi* (D'ORBIGNY) (1846, Foram. Foss. Bass. Tert. Vienne, p. 154, pl. 8, fig. 4–6).

Distribution. PHLEGER and PARKER (1951): maximum depth less than about 100 m., one record at 124 m.; frequencies less than 5 %. PARKER (1954): to a depth of 145 m., maximum frequency 9 % at 44 m. PHLEGER (1956): range 4–99 m., optimum 50–80 m. (both dead and living).

In our material frequencies are usually less than 2 %, except for two zones. One between 45 and 75 m. (max. freq. 7 %) in the eastern part of the Orinoco shelf, the other east of Trinidad between 25 and 110 m. (max. freq. 6 %). The highest frequencies are all between 45 and 70 m. The species seems to avoid dominantly pelitic areas.

Eponides regularis PHLEGER and PARKER

Map 16

Eponides regularis PHLEGER and PARKER, 1951, Geol. Soc. Am., Mem. 46, pt. 2, p. 21, pl. 11, fig. 3, 4; PARKER, 1954, Bull. Mus. Comp. Zool. Harv. Coll., vol. 111, no. 10, p. 529, pl. 9, fig. 16, 17.

Remarks. This is an easily distinguishable species in our material.

Distribution. In the Gulf of Mexico PHLEGER and PARKER (1951) found a depth range between 155 and 1430 m., PARKER (1954) from 145 to 3000 m. Maximum frequency 21 % at 200 metres. PHLEGER (1956) found living specimens between 70 and 101 metres.

In our material there is a marked area for this species in the deeper part of the mud platform north of Trinidad with the greatest frequency of 14 % at 135 metres. Frequencies of 5 % and more are all between 105 and 180 metres. Scattered low frequency occurrences were found on the Orinoco shelf, most of them near the slope.

Eponides repandus (FICHTEL and MOLL)

Pl. 2, fig. 8, 9

Nautilus repandus FICHTEL and MOLL, 1803, Test. Micr., p. 35, pl. 3, fig. a-d.

Eponides repandus (FICHTEL and MOLL), PHLEGER and PARKER, 1951, Geol. Soc. Am., Mem. 46, pt. 2, p. 21, pl. 11, fig. 5, 6; PARKER, 1954, Bull. Mus. Comp. Zool. Harv. Coll., vol. 111, no. 10, p. 529, pl. 9, fig. 27, 28.

Remarks. In our material the individuals may attain considerable size. Specimens referred to as *Poroeponides* cf. *lateralis* (TERQUEM) by PHLEGER and PARKER (1951, p. 23, pl. 12, fig. 5) have been included in *E. repandus.* Most of our larger specimens show a tendency to develop depressions and pores in the apertural face.

Distribution. Reported by PHLEGER and PARKER (1951) as a species occurring in calcareous areas with a depth range between 1 and 170 metres.

In our material scattered low frequency occurrences at all depths, usually avoiding pelitic areas. Higher frequencies (up to 6 %) were found between 20 and 40 m. east of Trinidad, and at scattered places off the mouth of the Essequibo. In the latter area they are partly found near the gully of the DK traverse. High frequencies at greater depths occur sporadically.

Eponides turgidus PHLEGER and PARKER

Eponides turgidus PHLEGER and PARKER, 1951, Geol. Soc. Am., Mem. 46, pt. 2, p. 22, pl. 11, fig. 9; PARKER, 1954, Bull. Mus. Comp. Zool. Harv. Coll., vol. 111, no. 10, p. 530, pl. 9, fig. 22, 23.

Distribution. In the Gulf of Mexico characteristic for depths greater than 100 m.; high frequencies deeper than 1500 m.

In our material most, low frequency, occurrences along the continental slope but occasional specimens as shallow as 17 metres. Highest frequency 3 %.

Gaudryina aequa Cushman

Pl. 2, fig. 10, 11

Gaudryina aequa Cushman, 1947, Contr. Cushm. Lab. Foram. Res., vol. 23, p. 87, pl. 18, fig. 18–21.

Gaudryina cf. *aequa* Cushman, Phleger and Parker, 1951, Geol. Soc. Am., Mem. 46, pt. 2, p. 6, pl. 2, fig. 11, 12; Parker, 1954, Bull. Mus. Comp. Zool. Harv. Coll., vol. 111, no. 10, p. 492, pl. 3, fig. 12.

Remarks. Parker (1951) remarked that her specimens were more angulate than the typical form, described by Cushman, but the paratypes of the species are said to show also more angulate borders. Our specimens are usually angular at the periphery; they very well resemble the figures given by Cushman and Parker. We found considerable variation in the relative length of the triserial part of the test. Some of the specimens with a long biseral portion usually have the periphery more rounded. Some others, usually those with long triserial part, have more or less distinct fistulose processes at the angles of the chambers.

Distribution. Phleger and Parker (1951) had the species at 7 stations between 50 and 75 metres; Parker (1954) found it down to a depth of 150 m.; Phleger (1956) recorded living specimens between 62 and 90 metres. Always low frequencies.

In our material scattered occurrences, always less than 2 %, between 50 and 130 metres.

Gaudryina atlantica (Bailey)

Pl. 2, fig. 12

Textularia atlantica Bailey, 1851, Smiths. Contr., vol. 2, art. 3, p. 12, fig. 38–43.

Gaudryina atlantica (Bailey), Cushman, 1937, Cushm. Lab. Foram. Res., Spec. Publ. no. 7, p. 95, pl. 14, fig. 4, 5; Phleger and Parker, 1951, Geol. Soc. Am., Mem. 46, pt. 2, p. 6, pl. 2, fig. 13.

Distribution. A few rare occurrences at 130 m. and deeper.

Gaudryina quadrangularis Bagg var. **antillana** Bermudez and Acosta

Pl. 3, fig. 1

Gaudryina quadrangularis Bagg var. *antillana* Bermudez and Acosta, 1940, Mem. Soc. Cub. Hist. Nat., vol. 14, p. 55, pl. 9, fig. 4, 5.

Remarks. Our specimens resemble this variety from the western Atlantic better than the figured specimens of the Pacific species *G. quadrangularis* Bagg (see Cushman, 1937, Cushm. Lab. Foram. Res., Spec. Publ. no. 7, p. 63, pl. 10, fig. 11, 15, 17).

Distribution. The type is from 290 m. off southern Cuba.

In our material rare from 50 to 150 m., with frequencies of 2 % or less; only at two stations shallower than 50 m.

Glandulina laevigata (D'ORBIGNY)

Nodosaria (Glandulina) laevigata D'ORBIGNY, 1826, Ann. Sci. Nat., vol. 7, p. 252, pl. 10, fig. 1–3.

Glandulina laevigata (D'ORBIGNY), CUSHMAN and OZAWA, 1930, Proc. U.S. Nat. Mus., vol. 77, art. 6, p. 143, pl. 40, fig. 1.

Remarks. Only three specimens.

Globobulimina affinis (D'ORBIGNY)

Bulimina affinis D'ORBIGNY, 1839, in Ramon de la Sagra, Hist. Fis. Pol. Nat. Cuba, Foraminifères, p. 105, pl. 2, fig. 25, 26; CUSHMAN and PARKER, 1947, U.S. Geol. Survey, Prof. Paper 210–D, p. 122, pl. 28, fig. 23–25; PHLEGER and PARKER, 1951, Geol. Soc. Am., Mem. 46, pt. 2, p. 15, pl. 7, fig. 21, 22.

Globobulimina affinis (D'ORBIGNY), PARKER, 1954, Bull. Mus. Comp. Zool. Harv. Coll., vol. 111, no. 10, p. 511, pl. 6, fig. 25, pl. 7, fig. 1.

Remarks. Only a few specimens.

Gypsina vesicularis (PARKER and JONES)

Orbitolina vesicularis PARKER and JONES, 1860, Ann. Mag. Nat. Hist., ser. 3, vol. 6, p. 31, no. 5.

Gypsina vesicularis (PARKER and JONES), H. B. BRADY, 1884, Rep. Voy. Challenger, Zoology, vol. 9, p. 718, pl. 101, fig. 9–12; PARKER, 1954, Bull. Mus. Comp. Zool. Harv. Coll., vol. 111, no. 10, p. 545, pl. 13, fig. 12.

Distribution. Scattered specimens at 8 stations of variable depth, between 20 and more than 183 metres.

Gyroidina cf. G. neosoldanii BROTZEN

Pl. 3, fig. 2

Cf. *Gyroidina neosoldanii* BROTZEN, 1936, Sver. Geol. Unders., ser. C, no. 396, p. 158.

Gyroidina neosoldanii BROTZEN, PARKER, 1954, Bull. Mus. Comp. Zool. Harv. Coll., vol. 111, no. 10, p. 528, pl. 9, fig. 9, 10.

Remarks. Nearly all our specimens are fairly small; they closely resemble the one figured by PARKER from the Gulf of Mexico. It is not clear whether they are really conspecific with BROTZEN's individuals, which have several more chambers in the final whorl (9–11) than our specimens (5–7). Some larger specimens resemble *G. soldanii* D'ORBIGNY var. *altiformis* R. E. and K. C. STEWART, as figured by PHLEGER and PARKER (1951, Geol. Soc. Am., Mem. 46, pt. 2, p. 22, pl. 11, fig. 15, 16) and PARKER (1954, p. 527, pl. 9, fig. 7, 8).

Distribution. In the Gulf of Mexico (PARKER) all occurrences are deeper than 183 metres, frequencies less than 1 %.

In our material scattered distribution at all depths, frequencies mostly less than 1 %. Frequencies of more than 2 % all deeper than 70 metres.

Hanzawaia concentrica (CUSHMAN)

Map 17

Truncatulina concentrica CUSHMAN, 1918, U.S. Geol. Survey, Bull. 676, p. 64, pl. 21, fig. 3.

Cibicides concentricus (CUSHMAN), CUSHMAN, 1930, Florida State Geol. Survey, Bull. no. 4, p. 61, pl. 12, fig. 4; CUSHMAN, 1931, U.S. Nat. Mus., Bull. 104, pt. 8, p. 120, pl. 21, fig. 4, 5, pl. 22, fig. 1, 2; PHLEGER and PARKER, 1951, Geol. Soc. Am., Mem. 46, pt. 2, p. 29, pl. 15, fig. 15.

Cibicidina concentrica (CUSHMAN), PARKER, 1954, Bull. Mus. Comp. Zool. Harv. Coll., vol. 111, no. 10, p. 544, pl. 13, fig. 7, 10.

Remarks. Our specimens commonly lack the concentrical dorsal band described by CUSHMAN (1918). They clearly belong to the *H. boueana* group. They differ from *H. boueana* (D'ORBIGNY) (1846, Foram. Foss. Bass. Tert. Vienne, p. 169, pl. 9, fig. 24–26; MARKS, 1951, Contr. Cushm. Found. Foram. Res., vol. 2, p. 72, pl. 8, fig. 24–26) in the number of chambers (7–9 instead of 9–11), in the periphery (subcarinate in *H. concentrica*, subacute to acute without a distinct keel in *H. boueana*) and in the flatness of the dorsal side (mostly flattened or slightly convex in *H. concentrica*, flattened or concave in *H. boueana*). The sutures are usually somewhat more limbate than in *H. boueana*, especially on the dorsal side, and the perforation is sometimes somewhat coarser.

Possibly part of the material described as *Cibicides americanus* (CUSHMAN) (1918, U.S. Nat. Mus., Bull. 103, p. 68, pl. 23, fig. 2.) belongs to the same species as our recent material, but the specimens from the Miocene of Aruba (DROOGER, 1953, Contr. Cushm. Found. Foram. Res., vol. 4, p. 145) show a hyaline keeled periphery, which is not present in our specimens.

Among our specimens the type of *H. concentrica* is dominant; individuals resembling *H. strattoni* (APPLIN) as figured from the Gulf of Mexico (PARKER, BANDY) are extremely rare.

Probably we included in our countings occasional specimens that are identical with *Discorbis bertheloti* (D'ORBIGNY) as figured by PHLEGER and PARKER (1951, pl. 10, fig. 1, 2) and PARKER (1954, pl. 8, fig. 22, 23).

We placed the species in the genus *Hanzawaia* ASANO, which is the oldest available name for this group of species, which are clearly different from *Cibicides*.

Distribution. In the Gulf of Mexico abundant (up to 35 %) in water less than 100 metres. PHLEGER (1956) found an abundant living population between 25 and 77 m., less frequent down to 90 m.

In our material rare in shallow pelitic areas. Highest frequencies (up to 25 %) between 35 and 75 m., but occasional high frequencies (up to 8 %) occur down to over 183 metres.

Heterostegina antillarum D'ORBIGNY.

Map 18

Heterostegina antillarum D'ORBIGNY, 1839, in Ramon de la Sagra, Hist. Fis. Pol. Nat. Cuba, Foraminifères, p. 121, pl. 7, fig. 24, 25; CUSHMAN, 1922, Carnegie Inst. Wash., Publ. 311, p. 57, pl. 10, fig. 5; CUSHMAN, 1930, U.S. Nat. Mus., Bull. 104, pt. 7, p. 33, pl. 12, fig. 1, 2.

Remarks. Larger specimens are close to *H. antillarum*, as interpreted by CUSHMAN; D'ORBIGNY's type figure is of little use. Evidently the involute coiling is reached only in the later ontogenetic stages of the test, since the smallest macrospheric individuals are evolute. Individuals of intermediate size resemble *H. suborbicularis* as figured by FORNASINI in 1904 (fig. 7, see BROOKS ELLIS and MESSINA).

Distribution. CUSHMAN's localities in the Dry Tortugas area are stated to be in abnormally warm water for the region.

East of Trinidad an area with up to 5% (at 45 m.) between 20 and 75 metres. North of the Essequibo some other occurrences at approximately 55 metres.

Höglundina elegans (D'ORBIGNY)

Map 19

Rotalia elegans D'ORBIGNY, 1826, Ann. Sci. Nat., vol. 7, p. 276.

Pulvinulina elegans (D'ORBIGNY), H. B. BRADY, 1884, Rep. Voy. Challenger, Zoology, vol. 9, p. 699, pl. 105, fig. 3–6.

Höglundina elegans (D'ORBIGNY), BROTZEN, 1948, Sver. Geol. Unders., ser. C, no. 493, Årsbok 42, no. 2, p. 92; PHLEGER and PARKER, 1951, Geol. Soc. Am., Mem. 46, pt. 2, p. 22, pl. 12, fig. 1; PARKER, 1954, Bull. Mus. Comp. Zool. Harv. Coll., vol. 111, no. 10, p. 531, pl. 10, fig. 4, 8.

Distribution. In the Gulf of Mexico (PHLEGER and PARKER, 1951) widespread and characteristic of many deep samples down to 3550 m., but the species also occurs in shallow water up to 90 m. According to PARKER (1954) widely distributed deeper than 65 m., 1–5 % deeper than 345 m. with a few higher frequencies deeper than 1300 metres. KRUIT (1954) reported the species rather common in the area of the Bocas del Dragon and concluded a real marine environment, largely below 90 metres.

In our material there is a very distinct high frequency area on the mud platform north of Trinidad and Paria between 50 and 140 m. The highest frequencies (up to 15 %) are found between 70 and 110 m. East of Trinidad occasional specimens were found along the continental slope.

Evidently depth is not the primary factor for the distribution of *H. elegans*. Its optimum conditions seem to be in muds of fully marine environment, probably independent of depth.

Karreriella bradyi (CUSHMAN)

Pl. 3, fig. 3

Gaudryina bradyi CUSHMAN, 1911, U.S. Nat. Mus., Bull. 71, pt. 2, p. 67, textf. 107.

Karreriella bradyi (CUSHMAN), CUSHMAN, 1937, Cushm. Lab. Foram. Res., Spec. Publ. no. 8, p. 135, pl. 16, fig. 6–11; PARKER, 1954, Bull. Mus. Comp. Zool. Harv. Coll., vol. 111, no. 10, p. 495, pl. 3, fig. 11.

Distribution. This species is rather characteristic of fairly deep water (CUSHMAN, 1937). Upper depth limit according to various authors between 90 and 155 metres.

In our material single specimens in only five samples, not shallower than 150 metres.

Lagena spp.

The *Lagena* individuals of our material belong to many species. Determinations have been based mainly on BRADY's Challenger Report. The species are occasionally taken in a wide sense.

Most common are *L. squamosa* (MONTAGU), *L. striata* (D'ORBIGNY) and *L. perlucida* (MONTAGU). Less frequent are *L. acuticosta* REUSS, *L. laevis* (MONTAGU) var. CUSHMAN and MCCULLOCH, *L. gracillima* (SEGUENZA), *L. striatopunctata* PARKER and JONES and *L. hispida* REUSS. Occasional specimens of still other species are not recorded.

Distribution. The group as a whole is represented in many samples at all depths, usually with low frequencies (maximally 3 %).

Lagena acuticosta REUSS

Lagena acuticosta REUSS, H. B. BRADY, 1884, Rep. Voy. Challenger, Zoology, vol. 9, p. 464, pl. 57, fig. 31, 32.

Distribution. DI 1046, DV 1147, DH 1040.

Lagena gracillima (SEGUENZA)

Lagena gracillima (SEGUENZA), H. B. BRADY, 1884, Rep. Voy. Challenger, Zoology, vol. 9, p. 456, pl. 56, fig. 21, 22, 24–26; CUSHMAN, 1923, U.S. Nat. Mus., Bull. 104, pt. 4, p. 23, pl. 4, fig. 5.

Distribution. DH 1040 and DQ 1103.

Lagena hispida REUSS

Lagena hispida REUSS, H. B. BRADY, 1884, Rep. Voy. Challenger, Zoology, vol. 9, p. 459, pl. 57, fig. 1–4.

Remarks. Some specimens resemble better BRADY's figures of *L. aspera* REUSS (pl. 57, fig. 7–12).

Distribution. DN 1082, DI 1054, DQ 1103, 291.

Lagena laevis (MONTAGU) var. CUSHMAN and MCCULLOCH

Lagena laevis (MONTAGU) var. CUSHMAN and MCCULLOCH, 1950, Allan Hancock Pac. Exp., vol. 6, no. 6, p. 342, pl. 45, fig. 18.

Remarks. Only the lower part of the test is spinose.

Distribution. DQ 1103 and DO 1089.

Lagena perlucida (Montagu)

Lagena perlucida (Montagu), Cushman, 1923, U.S. Nat. Mus., Bull. 104, pt. 4, p. 46, pl. 8, fig. 12, 13; Cushman and McCulloch, 1950, Allan Hancock Pac. Exp., vol. 6, no. 6, p. 343, pl. 46, fig. 1, 2.

Remarks. Various specimens with very fine striae, mainly on the aboral part of the test.

Distribution. In six samples at 70 m. and shallower.

Lagena squamosa (Montagu)

Lagena squamosa (Montagu), H. B. Brady, 1884, Rep. Voy. Challenger, Zoology, vol. 9, p. 471, pl. 58, fig. 28, 29; Cushman, 1923, U.S. Nat. Mus., Bull. 104, pt. 4, p. 51, pl. 10, fig. 3, 4; Drooger, 1953, Contr. Cushm. Found. Foram. Res., vol. 4, p. 125.

Remarks. All reticulate *Lagena* specimens have been gathered under the name *L. squamosa*. In addition to typical individuals, most others can be assigned to *L. squamosa* (Montagu) var. *hexagona* Williamson (1848, Ann. Mag. Nat. Hist., ser. 2, vol. 1, p. 20, pl. 2, fig. 23).

Distribution. Scattered low frequency occurrences (18 stations) at depths of 65 m. and deeper.

Lagena striata (d'Orbigny)

Lagena striata (d'Orbigny), H. B. Brady, 1884, Rep. Voy. Challenger, Zoology, vol. 9, p. 460, pl. 57, fig. 22, 24; Cushman, 1923, U.S. Nat. Mus., Bull. 104, pt. 4, p. 54, pl. 10, fig. 9; Cushman and McCulloch, 1950, Allan Hancock Pac. Exp., vol. 6, no. 6, p. 350, pl. 47, fig. 1–4.

Remarks. There are several finely striated specimens of variable shape. Some specimens tend towards *L. sulcata* (Walker and Jacob).

Distribution. Some 15 occurrences at all depths.

Lagena striatopunctata Parker and Jones

Lagena striatopunctata Parker and Jones, H. B. Brady, 1884, Rep. Voy. Challenger, Zoology, vol. 9, p. 468, pl. 58, fig. 37, 40; Cushman and McCulloch, 1950, Allan Hancock Pac. Exp., vol. 6, no. 6, p. 351, pl. 47, fig. 5–9.

Remarks. Some of our specimens closely resemble the figures of Cushman and McCulloch. These forms somewhat resemble *L. plumigera* H. B. Brady (1884, pl. 58, fig. 25, 27), but all have tubulated costae; in this respect they resemble *L. desmophora* Rymer-Jones (*L. vulgaris* Williamson var. *desmophora* Rymer-Jones, 1872, Trans. Linn. Soc. London, vol. 30, p. 54, pl. 19, fig. 23, 24), which is characterized by several longitudinal primary costae with tubulations, between which there are secondary costae that are solid and non-tubulated. However, we did not find specimens with such secondary costae.

Distribution. At 75 metres and shallower; DB 1003, 1005, DE 1021, EE 1210, 1214.

Lenticulina peregrina (Schwager)

Cristellaria peregrina Schwager, 1866, Novara Exp., Geol. Theil, vol. 2, p. 245, pl. 7, fig. 89.

Lenticulina peregrina (Schwager), Phleger and Parker, 1951, Geol. Soc. Am., Mem. 46, pt. 2, p. 9, pl. 4, fig. 19, 20; Parker, 1954, Bull. Mus. Comp. Zool. Harv. Coll., vol. 111, no. 10, p. 504, pl. 5, fig. 18.

Distribution. Rare individuals at some 7 stations with variable depth.

Liebusella soldanii (Jones and Parker)

Map 20

Haplostiche soldanii (Jones and Parker), H. B. Brady, 1884, Rep. Voy. Challenger, vol. 9, p. 318, pl. 32, fig. 12–18.

Haplostiche dubia, Cushman (not d'Orbigny), 1920, U.S. Nat. Mus., Bull. 104, pt. 2, p. 34, pl. 7, fig. 2, 3.

Liebusella soldanii (Jones and Parker), Cushman, 1937, Cushm. Lab. Foram. Res., Spec. Publ. no. 8, p. 166, pl. 20, fig. 1–11.

Remarks. Typical specimens occur frequently; they are of considerable size, especially at DG 1028. The average size of full-grown adults is highly variable, especially when the contents of different samples are compared. In several samples there is complete gradation to globular and ovoid juveniles in which the uniserial stage is incomplete or absent, and the radial partitions of adult uniserial chambers probably always lacking. Other samples only contain such small individuals, the features of which are usually indistinct from the exterior. The size range is again great, though fairly constant for single samples. Many of the immature specimens resemble *Goësella mississippiensis* Parker (1954, Bull. Mus. Comp. Zool. Harv. Coll., vol. 111, no. 10, p. 495, pl. 3, fig. 13, 14), a few others give the impression of having a similar arrangement as the much bigger *Ammosphaeroidina sphaeroidiniformis* (Brady), as described and figured by Cushman (1920, Bull. 104, pt. 2, p. 87, pl. 17, fig. 5). In the countings all small specimens have been referred to as *Liebusella* juvenile. They have been included on the map of *L. soldanii.*

Distribution. Cushman stated (1920) that *L. soldanii* is commonly associated with coral reef faunas, in comparatively shallow tropical waters. In the Gulf of Mexico Parker (1954) found some individuals with a depth range from 70 to 275 metres.

In our material there is an area with high frequencies (up to 18 %) off the Essequibo between 40 and 70 m. At a similar depth less high frequencies (up to 4 %) were found east of Trinidad. Other high frequencies occur on the deeper part of the Orinoco shelf with maxima of 20 % at DH 1028 (80 m.) and of 25 % at DZ 1173 on the continental slope. The species is absent from the pelite areas.

Loxostomum limbatum (Brady)

Pl. 3, fig. 4

Bolivina limbata H. B. Brady, 1884, Rep. Voy. Challenger, Zoology, vol. 9, p. 419, pl. 52, fig. 26–28.

Loxostoma limbatum (BRADY), CUSHMAN, 1937, Cushm. Lab. Foram. Res., Spec. Publ. no. 9, p. 186, pl. 21, fig. 26–29.

Remarks. A number of single specimens from several stations seem to agree best with this Pacific species, especially if the variety *costulatum* (CUSHMAN) (see CUSHMAN, 1937, id., p. 187, pl. 21, fig. 30, 31) is included. Most of them are immature specimens. Costae are variously developed. The group is possibly not homogeneous; affinities with *L. porrectum* (BRADY) and *L. mayori* (CUSHMAN) cannot be excluded.

Distribution. Scattered, rare occurrences; depth range from 25 to over 183 metres.

Marginulina subaculeata (CUSHMAN) var. **glabrata** (CUSHMAN)

Pl. 3, fig. 5

Cristellaria subaculeata CUSHMAN var. *glabrata* CUSHMAN, 1923, U.S. Nat. Mus., Bull. 104, pt. 4, p. 124, pl. 32, fig. 4, pl. 33, fig. 3, pl. 34, fig. 2, 3.

Marginulina subaculeata (CUSHMAN) var. *glabrata* (CUSHMAN), PHLEGER and PARKER, 1951, Geol. Soc. Am., Mem. 46, pt. 2, p. 9, pl. 5, fig. 4.

Remarks. Mainly juvenile specimens, from scattered stations. It was not possible to separate young specimens of this variety from those of *Vaginulina planata* (PHLEGER and PARKER).

Distribution. At all depths, mostly in samples deeper than 50 metres. Frequencies less than 1 %.

Miliolinella fichteliana (D'ORBIGNY)

Pl. 3, fig. 6

Triloculina fichteliana D'ORBIGNY, 1839, in Ramon de la Sagra, Hist. Fis. Pol. Nat. Cuba, Foraminifères, p. 171, pl. 9, fig. 8–10; CUSHMAN, 1929, U.S. Nat. Mus., Bull. 104, pt. 6, p. 63, pl. 17, fig. 1.

Distribution. Originally described from shore sands of Cuba and Jamaica.

In our material this species is confined to the surroundings of Trinidad and Tobago. Most occurrences are on the platform east of Trinidad with a maximal frequency of 4 % and a depth range of 20 to 65 metres. North of Trinidad there are some single individuals in deeper water (down to 200 m.), probably displaced. Occurring only in calcarenitic sediments.

Miliolinella subrotunda (MONTAGU)

Map 21

Vermiculum subrotundum MONTAGU, 1803, Test. Brit., pt. 2, p. 521.

Quinqueloculina subrotunda (MONTAGU), D'ORBIGNY, 1826, Ann. Sci. Nat., ser. 1, vol. 7, p. 302.

Triloculina subrotunda (MONTAGU), MARKS, 1951, Contr. Cushm. Found. Foram. Res., vol. 2, p. 40.

Triloculina circularis, CUSHMAN (not BORNEMANN?), 1929, U.S. Nat. Mus., Bull. 104, pt. 6, p. 58, pl. 13, fig. 6, 7, pl. 14, fig. 1, 2.

Remarks. The specimens of this species are very variable in shape and in arrangement of the chambers. Juvenile individuals, which are most abundant, are commonly quinqueloculine. The same variation is indicated in CUSHMAN's figures of *T. circularis.* This author suggested a constant form in the tropics but greater variability in colder waters. Possibly other factors are equally important in influencing variation. The same variation was found in the Miocene material of the Vienna (MARKS) and Aquitaine basins (KAASSCHIETER).

Because of the broad, flattened tooth the species is removed from the genus *Triloculina.*

Distribution. The species occurs in most of our samples, at variable depth. It is more or less distinctly absent from purely pelitic sediments. The pattern of the higher frequencies is not very clear.

Nodobaculariella cassis (D'ORBIGNY)

Vertebralina cassis D'ORBIGNY, 1839, in Ramon de la Sagra, Hist. Fis. Pol. Nat. Cuba, Foraminifères, p. 51, pl. 7, fig. 14, 15.

Nodobaculariella cassis (D'ORBIGNY), PHLEGER and PARKER, 1951, Geol. Soc. Am., Mem. 46, pt. 2, p. 8, pl. 4, fig. 12–14.

Distribution. Four occurrences between 48 and 183 metres.

Nodosaria intercellularis BRADY

Pl. 3, fig. 7

Nodosaria intercellularis H. B. BRADY, 1884, Rep. Voy. Challenger, Zoology, vol. 9, p. 515, pl. 65, fig. 1–4; CUSHMAN, 1923, U.S. Nat. Mus., Bull. 104, pt. 4, p. 89, pl. 14, fig. 2–4, pl. 17, fig. 3.

Remarks. Variation of our specimens approaches the types of *N. sublineata* BRADY (1884, p. 508, pl. 63, fig. 19–21). Possibly the forms referred to the latter species by PHLEGER and PARKER (1951, pt. 2, p. 10) belong to the same species as our *N. intercellularis.*

Distribution. Scattered distribution with low frequencies (less than 2 %) at all depths. Most occurrences are situated on the outer shelf below 75 metres.

Nodosaria pyrula D'ORBIGNY

Nodosaria pyrula D'ORBIGNY, 1826, Ann. Sci. Nat., vol. 7, p. 253, no. 13; H. B. BRADY, 1884, Rep. Voy. Challenger, Zoology, vol. 9, p. 497, pl. 62, fig. 10–12; CUSHMAN, 1923, U.S. Nat. Mus., Bull. 104, pt. 4, p. 69, pl. 16, fig. 1–4; PHLEGER and PARKER, 1951, Geol. Soc. Am., Mem. 46, pt. 2, p. 10, pl. 5, fig. 5.

Distribution. A few fragments in some 7 samples with wide depth range.

Nodosaria vertebralis (BATSCH)

Nautilus (Orthoceras) vertebralis BATSCH, 1791, Conchylien des Seesands, pt. 3, no. 6, pl. 2, fig. 6.

Dentalina vertebralis (BATSCH), CUSHMAN, 1931, Contr. Cushm. Lab. Foram. Res., vol. 7, p. 66, pl. 8, fig. 20. 21,

Nodosaria vertebralis (BATSCH), DROOGER, 1953, Contr. Cushm. Found. Foram. Res., vol. 4, p. 123.

Remarks. Our few specimens might as well be assigned to *N. vertebralis* (BATSCH) var. *albatrossi* CUSHMAN (1923, U.S. Nat. Mus., Bull. 104, pt. 4, p. 87, pl. 15, fig. 1).

Distribution. A few scattered occurrences of single specimens; 67 m. or deeper.

Nonion formosum (SEGUENZA)

Nonionina formosa SEGUENZA, 1880, Atti. R. Acc. Lincei, ser. 3, vol. 6, p. 63, pl. 7, fig. 6.

Nonion formosum (SEGUENZA), PARKER, 1954, Bull. Mus. Comp. Zool. Harv. Coll., vol. 111, no. 10, p. 506, pl. 6, fig. 3.

Remarks. The maximum number of chambers in the final whorl is 9 or 10 in our specimens, which may be somewhat lower than it is at other places.

Distribution. Frequencies 1 % or less; scattered occurrences at 30 m. and deeper.

Nonion grateloupi (D'ORBIGNY)

Nonionina grateloupi D'ORBIGNY, 1826, Ann. Sci. Nat., vol. 7, p. 294, no. 9; D'ORBIGNY, 1839, in Ramon de la Sagra, Hist. Fis. Pol. Nat. Cuba, Foraminifères, p. 46, pl. 6, fig. 6, 7.

Nonion grateloupi (D'ORBIGNY), CUSHMAN, 1939, U.S. Geol. Survey, Prof. Paper 191, p. 21, pl. 6, fig. 1–7; PHLEGER and PARKER, 1951, Geol. Soc. Am., Mem. 46, pt. 2, p. 11, pl. 5, fig. 18.

Remarks. Our rare specimens of *N. grateloupi* differ from those of *Nonionella atlantica* in the commonly smaller size, more compressed test and more elongate chambers. They are often slightly *Nonionella*-like. Possibly they are no more than variants of *Nonionella atlantica.*

Distribution. Our low frequency occurrences are shallower than about 125 m., most of them are found at less than 100 m.

Nonionella atlantica CUSHMAN

Map 22

Nonionella atlantica CUSHMAN, 1947, Contr. Cushm. Lab. Foram. Res., vol. 23, p. 90, pl. 20, fig. 4, 5; PHLEGER and PARKER, 1951, Geol. Soc. Am., Mem. 46, pt. 2, p. 11, pl. 5, fig. 21–23; BANDY, 1954, U.S. Geol. Survey, Prof. Paper 254–F, p. 137, pl. 29, fig. 10; KEY, 1954, Verh. K.N.A.W., afd. Natuurk., ser. 1, vol. 20, no. 5, p. 210, pl. 1, fig. 15; PARKER, 1954, Bull. Mus. Comp. Zool. Harv. Coll., vol. 111, no. 10, p. 507, pl. 6, fig. 6, 7.

Remarks. Typical specimens of considerable size were mainly found in the shallower samples, in which the species is most common. In the deeper samples the number of nonionid individuals decreases, while also their average size diminishes. These smaller specimens are generally somewhat different from *N. atlantica*, though it may be doubted whether they belong to really different species. Partly they are certainly variants of our most common *Nonionella* species. Some of them, resembling *Nonion sloani* (D'ORBIGNY), certainly belong to *Nonionella atlantica*. This is less certain for all our specimens of *Nonion grateloupi*.

Distribution. This is evidently a shallow water species that can stand some salinity decrease. PHLEGER and PARKER (1951) found it most abundant at less than 70 m. (8–10 %), but in quantity down to about 110 m., and with lower limit at 200 m. PARKER (1954) found frequencies up to 10 % shallower than 140 m., less than 1 % deeper than 140 m. Living specimens were reported as relatively abundant along the Texas coast by PHLEGER (1956) between 10 and 73 m.

In the Gulf of Paria (KRUIT, 1954), *N. atlantica* is one of the most abundant species. Together with *Rotalia rolshauseni* it characterizes the Gulf fauna. In the Serpents Mouth area it is the dominant species.

In our material *N. atlantica* is a very frequent species along the entire coast. Most higher frequencies are shallower than 75 m. Very high frequencies (up to 35 % at 58 m.) are found in the pelite area off the Orinoco. A band of highest frequencies in this area occurs between 30 and 50 m. and probably continues into the central area of the Serpents Mouth. Off the Essequibo the frequencies are less high (up to 20 %), the outer limit of the main distribution is shallower in this area, approximately at 40 m. Along the Trinidad-Paria coast frequencies are lower and more irregular, occasionally amounting up to 19 % near the Bocas del Dragon, in deeper water (123 m., possibly transported specimens).

Nonionella opima CUSHMAN

Textfig. 3

Nonionella opima CUSHMAN, 1947, Contr. Cushm. Lab. Foram. Res., vol. 23, p. 90, pl. 20, fig. 1–3; PARKER, 1954, Bull. Mus. Comp. Zool. Harv. Coll., vol. 111, no. 10, p. 507, pl. 6, fig. 10–12.

Nonionella cf. *opima* CUSHMAN, PHLEGER and PARKER, 1951, Geol. Soc. Am., Mem. 46, pt. 2, p. 11, pl. 6, fig. 1, 2.

Remarks. Our specimens have a wide range of variation in the size of the umbilical lobe. The same variation was described and figured for the specimens of the Gulf of Mexico. Some individuals are close to variants of *N. atlantica*.

Distribution. PHLEGER and PARKER (1951) reported this species common from shallowest depths sampled to about 100 m., highest frequencies (up to 9 %) from 30 to 70 m. PARKER (1954) found frequencies up to 20 % shallower than 100 m. (one occurrence of 31 % at 53 m.) and up to 5 % at greater depths. PHLEGER (1956) reported living specimens abundant between 7 and 79 m.

In our material there are scattered occurrences down to about 80 m. East of the Serpents Mouth, frequencies reach 6 % at about 40 m.

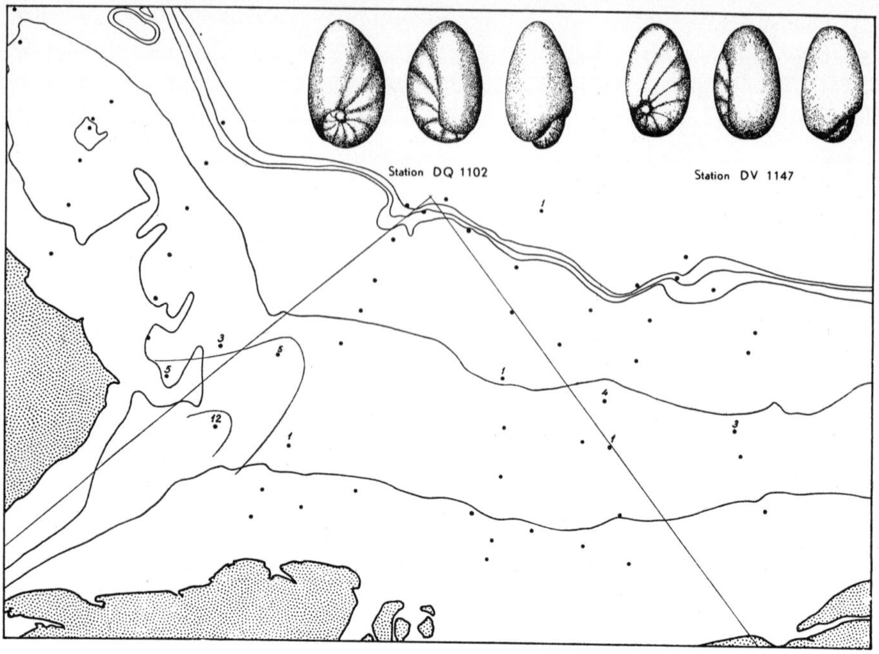

Fig. 3.
Distribution of *Nonionella opima* in part of the investigated area.
Figures ×65.

Patellina corrugata WILLIAMSON

Patellina corrugata WILLIAMSON, 1858, Rec. Foram. Gr. Brit., p. 146, pl. 3, fig. 86–89; PHLEGER and PARKER, 1951, Geol. Soc. Am., Mem. 46, pt. 2, p. 23, pl. 12, fig. 4.

Distribution. Only a few specimens.

Pavonina atlantica CUSHMAN

Pl. 3, fig. 8

Pavonina atlantica CUSHMAN, 1922, U.S. Nat. Mus., Bull. 104, pt. 3, p. 51, pl. 19, fig. 1; CUSHMAN, 1945, Contr. Cushm. Lab. Foram. Res., vol. 21, p. 48, pl. 8, fig. 11, 12; PHLEGER and PARKER, 1951, Geol. Soc. Am., Mem. 46, pt. 2, p. 17, pl. 8, fig. 6, 7.

Remarks. Our smaller specimens resemble best those of CUSHMAN's figures, the larger ones are closer to the individuals figured by PHLEGER and PARKER from the Gulf of Mexico. The latter are close to *P. miocenica* CUSHMAN and PONTON (see CUSHMAN, 1945, p. 46, pl. 8, fig. 1) from the Miocene of Florida, which species, in our opinion, is hardly different from *P. atlantica.*

Distribution. In our material there are low frequency occurrences, scattered over the entire area, mainly between 35 and 110 m., always in calcarenitic sediments.

Peneroplidae

Remarks. Specimens of the family Peneroplidae are very rare. They are usually worn, except for some of the smaller *Peneroplis* individuals. They belong to species of either *Peneroplis* or *Archaias.*

Distribution. There are a small number of occurrences scattered at all depths deeper than 45 metres. Frequencies are less than 1 %. Exception must be made for the occurrences at stations EA 1182 and 1183 (up to 3 %) at 46 and 22 metres east of Trinidad in the immediate vicinity of the klintite of Emerald Shoal.

Planorbulina mediterranensis D'ORBIGNY

Planorbulina mediterranensis D'ORBIGNY, 1826, Ann. Sci. Nat., vol. 7, p. 280, pl. 14, fig. 4–6; D'ORBIGNY, 1846, Foram. Foss. Bass. Tert. Vienne, p. 166, pl. 9, fig. 15–17; CUSHMAN, 1931, U.S. Nat. Mus., Bull. 104, pt. 8, p. 129, pl. 24, fig. 5–8; PHLEGER and PARKER, 1951, Geol. Soc. Am., Mem. 46, pt. 2, p. 33, pl. 19, fig. 5; PARKER, 1954, Bull. Mus. Comp. Zool. Harv. Coll., vol. 111, no. 10, p. 545, pl. 13, fig. 9.

Remarks. Young specimens of our *P. mediterranensis* resemble *Cibicides lobatulus* (WALKER and JACOB) var. *variabilis* (D'ORBIGNY) (*Truncatulina variabilis* D'ORBIGNY, 1839, in Parker, Webb and Berthelot, Hist. Nat. Isles Canaries, vol. 2, pt. 2, Foraminifères, p. 135, pl. 2, fig. 29). In larger specimens the double aperture per chamber, characteristic of *Planorbulina*, is clearly visible.

Distribution. PHLEGER and PARKER (1951) report a depth range from about 20 to 200 m., greatest frequencies between 40 and 100 m. PARKER stated that the species occurs mainly to a depth of 185 m.

In our material scattered occurrences, mainly in coarser sediments. Frequencies up to 2 %. Depth range 20 to 200 m., most occurrences between 45 and 125 m.

Planulina foveolata (BRADY)

Map 23

Anomalina foveolata H. B. BRADY, 1884, Rep. Voy. Challenger, Zoology, vol. 9. p. 674, pl. 94, fig. 1.

Planulina foveolata (BRADY), CUSHMAN, 1931, U.S. Nat. Mus., Bull. 104, pt. 8, p. 111, pl. 20, fig. 2, 3; PHLEGER and PARKER, 1951, Geol. Soc. Am., Mem. 46, pt. 2, p. 33, pl. 18, fig. 9, 10; PARKER, 1954, Bull. Mus. Comp. Zool. Harv. Coll., vol. 111, no. 10, p. 540, pl. 11, fig. 25, 26.

Remarks. Among small specimens there are some tendencies towards *P. exorna* PHLEGER and PARKER (1951, pt. 2, p. 32, pl. 18, fig. 5, 6).

Occasional individuals seem to be closer to *P. ariminensis* D'ORBIGNY (see PHLEGER and PARKER, 1951, pt. 2, p. 32, pl. 18, fig. 4) than to *P. foveolata*.

Distribution. In the Gulf of Mexico PHLEGER and PARKER (1951) give a range for the greatest frequencies from 120 to 200 m.; not present shallower than 85 m. PARKER (1954) reports a range from 75 to 530 m.; between 120 and 270 m. frequencies may be as high as 5 %.

In our material *P. foveolata* is mainly found along the border of the shelf. The shallowest record is at 67 m. All frequencies of 2 % and more are deeper than 120 m. The maximal frequency is 15 % at 293 at a depth of 166 m. north of the Bocas del Dragon.

The occasional specimens of *P. ariminensis* were found at 65 m. and deeper.

Proteonina spp.

Remarks. Irregularly scattered in our samples we found a number of coarsely arenaceous individuals of variable size and shape.

Pseudoclavulina mexicana (CUSHMAN)

Pl. 3, fig. 9

Clavulina humilis BRADY var. *mexicana* CUSHMAN, 1922, U.S. Nat. Mus., Bull. 104, pt. 3, p. 83, pl. 16, fig. 1–3.

Pseudoclavulina mexicana (CUSHMAN), CUSHMAN, 1937, Cushm. Lab. Foram. Res., Spec. Publ. no. 7, p. 117, pl. 16, fig. 5–11; PHLEGER and PARKER, 1951, Geol. Soc. Am., Mem. 46, pt. 2, p. 6, pl. 2, fig. 14–16; PARKER, 1954, Bull. Mus. Comp. Zool. Harv. Coll., vol. 111, no. 10, p. 493, pl. 3, fig. 8.

Remarks. Distinct specimens of *P. mexicana* show considerable variation in the absolute dimensions of the test and, often correspondingly, in the roughness of the arenaceous material of the wall. This variation is striking when specimens from different stations are compared; in a single sample the specimens are usually fairly constant.

A few individuals were found that are different by the usually smaller size, shorter and not or indistinctly triangular early portion, and relative smoothness of the wall. Their specific and even generic identity are uncertain. Possibly they belong to *P. novangliae* (CUSHMAN) or *P. curta* CUSHMAN and BRONNIMANN, or to *Martinottiella communis* (D'ORBIGNY). These rare, doubtful specimens commonly occur together with distinct *P. mexicana.*

Distribution. Scattered occurrences (2 % or less) with upper depth limit at about 50 m.

Pullenia bulloides (D'ORBIGNY)

Nonionina bulloides D'ORBIGNY, 1846, Foram. Foss. Bass. Tert. Vienne, p. 107, pl. 5, fig. 9, 10.

Pullenia bulloides (D'ORBIGNY), PHLEGER and PARKER, 1951, Geol. Soc. Am., Mem. 46, pt. 2, p. 29, pl. 15, fig. 11; CUSHMAN and TODD, 1942, Contr. Cushm. Lab. Foram. Res., vol. 18, p. 13, pl. 2, fig. 15–18; PARKER, 1954, Bull. Mus. Comp. Zool. Harv. Coll., vol. 111, no. 10, p. 538, pl. 11, fig. 17.

Remarks. Most of our specimens are slightly compressed.

Distribution. Frequencies 1 % or less from about 60 m. to over 180 m.

Pullenia quinqueloba (REUSS)

Nonionina quinqueloba REUSS, 1851, Zeitschr. d. d. geol. Ges., vol. 3, p. 71, pl. 5, fig. 3.

Pullenia quinqueloba (REUSS), PHLEGER and PARKER, 1951, Geol. Soc. Am., Mem. 46, pt. 2, p. 29, pl. 15, fig. 12, 13; PARKER, 1954, Bull. Mus. Comp. Zool. Harv. Coll., vol. 111, no. 10, p. 538, pl. 11, fig. 16.

Remarks. Most of our specimens have but four visible chambers.

Distribution. Several low frequency occurrences from 60 m. downwards.

Pyrgo fornasinii CHAPMAN and PARR

Biloculina ringens, H. B. BRADY (not *Miliolites ringens* LAMARCK), 1884, Rep. Voy. Challenger, Zoology, vol. 9, p. 142, pl. 2, fig. 7.

Pyrgo fornasinii CHAPMAN and PARR, 1935, Jour. Roy. Soc. W. Austr., vol. 21, p. 5.

Remarks. Our specimens may attain considerable size. The tooth is variable in width; generally it is broad. The individuals determined as *P. murrhyna* (SCHWAGER) by PHLEGER and PARKER (1951, pl. 3, fig. 11) and PARKER (1954, pl. 5, fig. 7) may be conspecific with our specimens.

Distribution. Scattered occurrences, low frequencies, at 45 m. and deeper; most records on the outer part of the shelf.

Pyrgo nasuta CUSHMAN

Pyrgo nasuta CUSHMAN, 1935, Smiths. Misc. Coll., vol. 91, no. 21, p. 7, pl. 3, fig. 1–4; PHLEGER and PARKER, 1951, Geol. Soc. Am., Mem. 46, pt. 2, p. 7, pl. 3, fig. 12–14.

Remarks. Typical specimens were found to grade into variants that resemble the individuals figured by PHLEGER and PARKER.

Distribution. Low frequencies (up to 3 %) at about 40 m. and deeper.

Quinqueloculina spp.

Map 24

The distribution of all our *Quinqueloculina* individuals gives a fairly uniform pattern with very high frequencies in the sandy areas. The highest ones occur in the shallow area immediately east of Trinidad (up to 37 %). Frequencies gradually decrease towards the border of the shelf. High frequencies are furthermore found on the Orinoco shelf, mainly on its outer part, and in the Cumberland Bank area.

Quinqueloculina compta CUSHMAN

Quinqueloculina compta CUSHMAN, 1947, Contr. Cushm. Lab. Foram. Res., vol. 23, p. 87, pl. 19, fig. 2; PHLEGER and PARKER, 1951, Geol. Soc. Am., Mem. 46, pt. 2, p. 7, pl. 3, fig. 16, 17; PARKER, PHLEGER and PEIRSON, 1953, Cushm. Found. Foram. Res., Spec. Publ. no. 2, p. 12, pl. 2, fig. 5, 6; BANDY, 1954, U.S. Geol. Survey, Prof. Paper 254–F, p. 138, pl. 28, fig. 2.

Remarks. Typical angular specimens, as originally figured by CUSHMAN (1947), are frequent among our material. Later authors (PARKER, BANDY) gave the species a much wider sense. During our counting the species was given a similar wide morphological range. Nearly all individuals are relatively small. It is quite possible that young specimens of other *Quinqueloculina* species have been included. It appeared impossible to make a clear distinction between the various forms.

Distribution. PARKER (1954) found frequencies up to 3 % at less than 50 m.; deeper, down to 155 m., percentages were smaller. PHLEGER (1956) reported living specimens at 10 stations between 11 and 30 metres.

In our material scattered at all depths, frequencies higher than 4 % (up to 13 %) shallower than 75 metres. Our data may be unreliable because of mixing of different species.

Quinqueloculina horrida CUSHMAN

Quinqueloculina horrida CUSHMAN, 1947, Contr. Cushm. Lab. Foram. Res., vol. 23, p. 88, pl. 19, fig. 1; PHLEGER and PARKER, 1951, Geol. Soc. Am., Mem. 46, pt. 2, p. 7, pl. 3, fig. 18, 19; PARKER, 1954, Bull. Mus. Comp. Zool. Harv. Coll., vol. 111, no. 10, p. 497, pl. 4, fig. 3, 4.

Remarks. Some specimens of arenaceous *Sigmoilina* species may have been incorporated.

Distribution. According to PARKER (1954) down to a depth of 370 metres, and frequencies less than 3 %. KRUIT reported the species from the Bocas del Dragon and the Serpents Mouth, commonly not deeper than 40 m.

In our material scattered, low frequency occurrences. Some concentrations were found between 35 and 75 metres, and a number of others along the continental slope, at variable, but greater depth.

Quinqueloculina lamarckiana D'ORBIGNY

Map 25

Quinqueloculina lamarckiana D'ORBIGNY, 1839, in Ramon de la Sagra, Hist. Fis. Pol. Nat. Cuba, Foraminifères, p. 189, pl. 11, fig. 14, 15; CUSHMAN, 1929, U.S. Nat. Mus., Bull. 104, pt. 6, p. 26, pl. 2, fig. 6; PHLEGER and PARKER, 1951, Geol. Soc. Am., Mem. 46, pt. 2, p. 7, pl. 4, fig. 1.

Remarks. Not all our specimens show an acute periphery; sometimes the periphery is very slightly rounded. All individuals with at least one edge with two carinae were determined as *Q. lamarckiana* var. *bicostata.*

Distribution. According to PARKER (1954) down to 275 metres, but frequencies are highest (1–5 %) at the shoaler ends of the traverses. In the Gulf of Paria the species was found to be common to abundant in the sandy areas of Serpents Mouth, Soldado Rock, Bocas del Dragon and Guiria trough. Some salinity decrease is tolerated.

In our material the species was combined with its variant *Q. lamarckiana* var. *bicostata*, which usually attains still greater size. They are most frequent (up to 36 %) in the shallow sandy area off eastern Trinidad with the greatest frequencies shallower than 40 m. Evidently there is a relation with the coarser sediments, since a band of higher frequencies is found on the outer shelf, more or less avoiding the pelitic area off the Orinoco, but turning shorewards in the area of the Essequibo delta. West of Tobago higher frequencies are found again in the area with coarser sediments north of the Paria peninsula.

Quinqueloculina lamarckiana D'ORBIGNY var. bicostata D'ORBIGNY

Map 25

Quinqueloculina bicostata D'ORBIGNY, 1839, in Ramon de la Sagra, Hist. Fis. Pol. Nat. Cuba, Foraminifères, p. 195, pl. 12, fig. 8–10; PHLEGER and PARKER, 1951, Geol. Soc. Am., Mem. 46, pt. 2, p. 7, pl. 3, fig. 15; PARKER, 1954, Bull. Mus. Comp. Zool. Harv. Coll., vol. 111, no. 10, p. 496, pl. 4, fig. 1, 2.

Remarks. Our specimens are considered to be variants of *Q. lamarckiana.* There are various transitions in the character of the periphery: from simply acute to acute with two carinae.

Distribution. PARKER reported this form down to 145 m., shallower than 50 m. frequencies may amount to 3 %.

High frequencies in our material go together with high frequencies of the species.

Quinqueloculina vulgaris D'ORBIGNY

Quinqueloculina vulgaris D'ORBIGNY, 1826, Ann. Sci. Nat., vol. 7, p. 302, no. 33; CUSHMAN, 1929, U.S. Nat. Mus., Bull. 104, pt. 6, p. 25, pl. 2, fig. 3; PHLEGER and PARKER, 1951, Geol. Soc. Am., Mem. 46, pt. 2, p. 8, pl. 4, fig. 2.

Remarks. The larger individuals fit within the range of variation indicated by PHLEGER and PARKER. Small individuals usually have a flap-like tooth, instead of the narrower tooth of larger specimens. It is uncertain whether they should be assigned to a separate species. Except for the shape of the tooth both types are identical. Small individuals of *Miliolinella subrotunda* may have occasionally been included in the countings.

Distribution. In our material there are low frequencies and scattered occurrences between 40 and 110 metres.

Rectoglandulina comatula (CUSHMAN)

Nodosaria comatula CUSHMAN, 1923, U.S. Nat. Mus., Bull. 104, pt. 4, p. 83, pl. 14, fig. 5.

Pseudoglandulina comatula (CUSHMAN), PHLEGER and PARKER, 1951, Geol. Soc. Am., Mem. 46, pt. 2, p. 10, pl. 5, fig. 7–9; PARKER, 1954, Bull. Mus. Comp. Zool. Harv. Coll., vol. 111, no. 10, p. 505, pl. 5, fig. 22.

Remarks. There are no indications of an early biserial arrangement.

Rectoglandulina replaces *Pseudoglandulina*, the type species of which was found to belong to *Nodosaria* (LOEBLICH and TAPPAN, 1955, Smiths. Misc. Coll., vol. 126, no. 3, publ. 4189).

Distribution. Some single specimens with the shallowest occurrence at 128 m.

Reussella atlantica CUSHMAN

Map 26

Reussella spinulosa (REUSS) var. *atlantica* CUSHMAN, 1947, Contr. Cushm. Lab. Foram. Res., vol. 23, p. 91, pl. 20, fig. 6, 7.

Reussella atlantica CUSHMAN, PHLEGER and PARKER, 1946, Geol. Soc. Am., Mem.

46, pt. 2, p. 18, pl. 8, fig. 8, 9; PARKER, 1954, Bull. Mus. Comp. Zool. Harv. Coll., vol. 111, no. 10, p. 519, pl. 7, fig. 28.

Remarks. Our specimens are referred to as *R. atlantica* for a more obvious comparison with the data from the Gulf of Mexico. It is the most important spinose *Reussella* species described from recent deposits of the western Atlantic.

Systematics of the Miocene to recent spinose Reussellae is far from clear. *R. atlantica* is said to differ from the Miocene *R. spinulosa* (REUSS) "in the smaller size, more slender form, thinner-walled and more coarsely perforated test, and tendency for the later portion in the adult to decrease in diameter" (CUSHMAN, 1947). From the Miocene have been described *R. spinulosa* (REUSS) (type region Vienna basin; sutures level with the surface), *R. pulchra* CUSHMAN (type region Vienna basin; raised sutures and often finely spinose) and *R. spinulosa* var. *laevigata* CUSHMAN (type region Aquitaine basin; lack of spinose angles). A large suite of specimens of these taxonomic units were available from the Miocene of the Vienna basin (see MARKS, 1951, Contr. Cushm. Found. Foram. Res., vol. 2, p. 61). They belong to a single series of intergrading types, evidently representing a single species with a rather wide range of variation.

A similar variation was observed among our recent specimens, which differ from the Miocene series only in the smaller average size and thinner wall, which latter difference may be due to the fossilized state of the Miocene individuals.

Distribution. In the Gulf of Mexico the species is most common, up to 10 %, down to a depth of 120 m (PHLEGER and PARKER, 1951). According to PARKER (1954) down to 235 m., maximum of 8 % at 71 m. PHLEGER (1956) found the species most common between 30 and 110 m.

In our material there is some correlation between the abundance of the species and the coarse calcarenitic and sandy areas. Most high frequencies are between 40 and 80 metres.

Reussella minuta sp. nov.

Pl. 3, fig. 10

Description. Test about twice as long as broad, triangular in transverse section, gradually tapering towards the bluntly pointed initial end, with nearly parallel margins near the apertural end in adult specimens. Chambers triserially arranged, sharply triangular in shape, gradually increasing in size as added, with flattened sides and overhanging lower margins, and with more or less distinctly raised rims along the lower and vertical edges. Sutures straight and oblique, indistinct in the early portion of the test. Wall of the chambers smooth, finely perforate, with a series of beads or small spines along the overhanging lower margins and with a projecting point at the angle of each chamber. This ornamentation gives a ragged appearance to the early part of the test. Aperture wide, occupying nearly the entire apertural face and surrounded by a thickened rim; the lower end is drawn out to a point that touches the inner margin of the last-formed chamber. Observed maximum length 0.35 mm. Coll. no. S 6109, 6110.

Remarks. Only *R. checchia-rispolii* KICINSKY (1952, Contr. Sci. Geol., Cons. Naz. Ricerche, Centro di Studio Geol. Ital. Centro-Merid., vol. 2, p. 34, pl. 2, fig. 3–5) from the Lower Pleistocene of Central Italy is close to our *R. minuta*, from which it differs mainly by the absence of surface ornamentation.

Distribution. This minute species is rare (1 % or less) at several stations, scattered between 35 and 200 metres.

Reophax spp.

Remarks. Few indeterminable specimens and fragments in a small number of samples.

Robertina bradyi CUSHMAN and PARKER

Robertina bradyi CUSHMAN and PARKER, 1936, Contr. Cushm. Lab. Foram. Res., vol. 12, p. 99, pl. 16, fig. 9; CUSHMAN and PARKER, 1947, U.S. Geol. Survey, Prof. Paper 210–D, p. 75, pl. 18, fig. 16; PARKER, 1954, Bull. Mus. Comp. Zool. Harv. Coll., vol. 111, no. 10, p. 510, pl. 6, fig. 18.

Distribution. Only single specimens at three places on the shelf east of Trinidad, at 82 m. and deeper.

Robulus spp.

Map 27

Remarks. Very great variation of size is one of the reasons that the majority of our *Robulus* individuals cannot be separated into distinct species. In the countings they have been taken together. Because of the spines and the wide flange *R. calcar* and *R. iotus* could be fairly successfully distinguished. Furthermore there are among our material specimens that closely resemble the types of the following species: *R. cushmani*, *R. orbicularis*, *R. occidentalis* and *R. vortex.* Most individuals are in one way or other intermediate between these various forms. At one station (DZ 1173) the specimens were referable to *R. septentrionalis. R. occidentalis* and *R. orbicularis* forms are most common.

Distribution. From the Gulf of Mexico scattered occurrences have been reported. According to PHLEGER and PARKER (1951) the genus is probably characteristic of depths less than about 400 or 500 metres.

In our material the group also shows a scattered occurrence at all depths, but it is most common in the deeper parts of the traverses, frequencies varying up to 11 %. Those higher than 2 % occur at depths of 80 m. or more.

Robulus calcar (LINNÉ)

Nautilus calcar LINNÉ, 1767, Syst. Nat., 12th ed., p. 1162, no. 272.

Cristellaria calcar (LINNÉ), H. B. BRADY, 1884, Rep. Voy. Challenger, Zoology, vol. 9, p. 551, pl. 70, fig. 9–12, 15.

Robulus calcar (LINNÉ), CUSHMAN, 1923, U.S. Nat. Mus., Bull. 104, pt. 4, p. 115, pl. 30, fig. 7, pl. 31, fig. 4, 5.

Remarks. Specimens without spines would be inseparable from *R. cushmani* GALLOWAY and WISSLER.

Distribution. Scattered at depths of 80–200 metres.

Robulus cushmani GALLOWAY and WISSLER

Cristellaria rotulata, CUSHMAN (not LAMARCK), 1923, U.S. Nat. Mus., Bull. 104, pt. 4, p. 108, pl. 28, fig. 1, 2.

Robulus cushmani GALLOWAY and WISSLER, 1927, Jour. Pal., vol. 1, p. 51, pl. 8, fig. 11.

Distribution. Scattered occurrences, shallowest at 52 m.

Robulus iotus (CUSHMAN)

Cristellaria iota CUSHMAN, 1923, U.S. Nat. Mus., Bull. 104, pt. 4, p. 111, pl. 29, fig. 2, pl. 30, fig. 1.

Distribution. Only rare in a few samples, shallowest at 82 m.

Robulus occidentalis (CUSHMAN)

Cristellaria occidentalis CUSHMAN, 1923, U.S. Nat. Mus., Bull. 104, pt. 4, p. 102, pl. 25, fig. 2, pl. 26, fig. 1, 2.

Remarks. Included in this species are scattered individuals determined by CUSHMAN (1923) as *Cristellaria gibba* (D'ORBIGNY) and *C. d'orbignii* (BAILEY.)

Distribution. Scattered occurrences, shallowest at 52 m.

Robulus orbicularis (D'ORBIGNY)

Robulina orbicularis D'ORBIGNY, 1826, Ann. Sci. Nat., vol. 7, p. 288, pl. 15, fig. 8, 9.

Cristellaria orbicularis (D'ORBIGNY), H. B. BRADY, 1884, Rep. Voy. Challenger, Zoology, vol. 9, p. 549, pl. 69, fig. 17; CUSHMAN, 1923, U.S. Nat. Mus., Bull. 104, pt. 4, p. 101, pl. 21, fig. 7.

Remarks. Most of our specimens, referred to this species, are within the range of variation figured by BRADY and CUSHMAN. It is not possible to make a sharp distinction between this species and *R. cushmani* GALLOWAY and WISSLER, and between this species and *R. vortex* or some of our other types.

Distribution. Scattered at all depths. Frequencies low.

Robulus septentrionalis (CUSHMAN)

Cristellaria septentrionalis CUSHMAN, 1923, U.S. Nat. Mus., Bull. 104, pt. 4, p. 107, pl. 27, fig. 1, 2.

Remarks. At one station only, DZ 1173. Brown, possibly fossil specimens at 183 metres.

Robulus vortex (Fichtel and Moll)

Nautilus vortex Fichtel and Moll, 1803, Test. Micr., p. 33, pl. 2, fig. d-i.
Cristellaria vortex (Fichtel and Moll), H. B. Brady, 1884, Rep. Voy. Challenger, Zoology, vol. 9, p. 548, pl. 69, fig. 14–16; Cushman, 1923, U.S. Nat. Mus., Bull. 104, pt. 4, p. 108.

Distribution. Scattered at all depths. Shallowest occurrence at 50 metres.

The genus *Rotalia* is taken in a wide, conservative sense. Some of the species, such as *R. beccarii*, might be assigned to the genus *Streblus*, but in breaking up the genus *Rotalia* s.l. the position of some of the other species would be uncertain.

Rotalia bassleri Cushman and Cahill

Map 28

Rotalia bassleri Cushman and Cahill, 1933, U.S. Geol. Survey, Prof. Paper 175–A, p. 32, pl. 11, fig. 2.

Remarks. Our specimens show a fair degree of resemblance with this species from the Miocene of Maryland. *R. versiformis* Bandy (1953, Jour. Pal., vol. 27, p. 179, pl. 22, fig. 5) from recent deposits near California, is different by its raised dorsal sutures. In our material *R. bassleri* is perfectly separable from the other *Rotalia* species. Some variation was found in the size of the umbilical plug, which is, however, always distinct. The peripheral margin of the last two or three chambers is sometimes slightly lobulate.

The lack of peripheral views in Parker's publications unfortunately hampers a comparison with the Gulf of Mexico species.

Distribution. The map of this species shows a scattered low frequency distribution with remarkable concentrations off the mouth of the Essequibo, and along the northeastern coast of Trinidad (frequencies up to 9 and 7 %), both shallower than about 40 metres.

Rotalia beccarii (Linné) vars.

Map 29

Nautilus beccarii Linné, 1758, Syst. Nat., ed. 10, p. 710.

Remarks. Nearly all our specimens are relatively small for the species. They show considerable variation, which includes *R. beccarii* var. *tepida* Cushman and var. *parkinsoniana* (d'Orbigny), as figured by Phleger and Parker (1951, pt. 2, pl. 12, fig. 6, 7) from the Gulf of Mexico.

Distribution. In the Gulf of Mexico the species, i.e. mainly its variants, are reported as abundant to common in the shallow parts of the investigated areas, often more than 50 % of the entire fauna. Phleger (1956) found both living and dead specimens abundant between 4 and 37 m., and common between 38 and 75 m. Also Phleger and Parker (1951), and Parker (1954) report greatest frequencies down to 70 m. Kruit found the species common to abundant in the Serpents Mouth area, especially towards the coast of the Orinoco delta.

In our material very high frequencies are found in the coastal part of the Orinoco shelf in water shallower than 40 m., the contours fairly well following the 20 fathom line. Percentages increase shorewards (up to 54 %). This zone evidently continues in that noted by KRUIT in the Serpents Mouth area. Along the Trinidad and Paria coasts the frequencies are very much smaller. There is little doubt that salinity fluctuations of considerable amount favour the relative importance of this species.

Rotalia pauciloculata PHLEGER and PARKER

Map 30

Rotalia pauciloculata PHLEGER and PARKER, 1951, Geol. Soc. Am., Mem. 46, pt. 2, p. 23, pl. 12, fig. 8, 9.

Remarks. Probably our interpretation of this species is not fully correct, but it cannot be decided because of the insufficient figures given by PHLEGER and PARKER. All our specimens show about the same number of chambers, the narrow dorsal strips of these chambers, and the roughened ventral area around the umbilicus, which is often a distinct pit. These features fit in very well with those given by PHLEGER and PARKER. However, the ventral side is usually more convex than the dorsal one, and most specimens have a distinct peripheral keel, which latter feature seems to be absent in the original specimens. Possibly our keeled specimens may include immature *R. rolshauseni*, but in all other features they are closer to the remainder of our *R. pauciloculata* specimens.

Distribution. According to PHLEGER and PARKER this species is characteristic of water shallower than about 75 m., frequencies up to 5 %. PHLEGER (1956) found living specimens in several samples between 15 and 42 m., and at five stations between 44 and 75 metres. Frequencies up to 4 %.

In our material the distribution of this species fits in fairly well with that of *R. rolshauseni*. Both species often occur together or their areas are adjoining. *R. pauciloculata* is mainly found in pelitic sediments; frequencies up to 6 %; no occurrences deeper than 75 metres.

Rotalia rolshauseni CUSHMAN and BERMUDEZ

Map 30

Rotalia rolshauseni CUSHMAN and BERMUDEZ, 1946, Contr. Cushm. Lab. Foram. Res., vol. 22, p. 119, pl. 19, fig. 11–13; PHLEGER and PARKER, 1951, Geol. Soc. Am., Mem. 46, pt. 2, p. 23, pl.12, fig. 10; KEY, 1954, Verh. K.N.A.W., ser. 1, vol. 20, no. 5, p. 215, pl. 2, fig. 14.

Rolshausenia rolshauseni (CUSHMAN and BERMUDEZ), BERMUDEZ, 1952, Bol. Geol. Venez., vol. 2, no. 4, p. 63, pl. 9, fig. 8.

Remarks. Distinct larger specimens are frequent, but it is very remarkable that small specimens are extremely rare. Possibly they have been included in other species, such as *R. pauciloculata*.

Distribution. In the Gulf of Mexico authors give restricted depth ranges. PHLEGER and PARKER (1951): between 24 and 48 m., BANDY (1954): down to 24 m., PHLEGER (1956): dead specimens 17 to 37 m., living ones between 14 and 27 m. In the Gulf of Paria the species is very frequent, except for the Bocas del Dragon area, i.e. down to about 35 m. Along the southern side of the Serpents Mouth the species continues less distinctly at the outer side of the *R. beccarii* zone (KRUIT, 1954).

In our material we found a number of *R. rolshauseni* occurrences off the mouth of the Orinoco along the outer side of the main area of *R. beccarii*. This is evidently the continuation of the conditions found by KRUIT in the Serpents Mouth. These occurrences are in pelite. Further eastwards the strip of occurrences is getting into shallower water. Off the Essequibo mouth it lies in less pelitic sediments and in the interior part of the *R. beccarii* distributional area.

Combination of these data enables the assumption that the relative abundance of *R. rolshauseni* depends of salinity fluctuations that are smaller than those of *R. beccarii*. The dominant *Rotalia* species in the outer part of the *R. beccarii* area of the Essequibo delta thus would be of subrecent age. However, a greater preference of *R. rolshauseni* for more pelitic sediments as the dominant factor for its distribution cannot be entirely excluded.

Frequencies in our material amount up to 8 %, the greatest depth is at about 60 metres.

Rotalia sarmientoi REDMOND

Map 31

Rotalia sarmientoi REDMOND, 1953, Jour. Pal., vol. 27, p. 726, pl. 76, fig. 11.

Rotalia beccarii (LINNÉ) var. A. KEY, 1954, Verh. K.N.A.W., ser. 1, vol. 20, no. 5, p. 214, pl. 2, fig. 13.

Remarks. The dorsal sutures of this species are commonly limbate and raised; in some specimens they are only limbate. The plug is somewhat variable in size, but distinct in all individuals. There is a fair agreement of our individuals with description and figure of REDMOND's Miocene species. Generally there are no difficulties in separating the specimens of *R. sarmientoi* from the accompanying individuals of *R. beccarii*: the dorsal sutures are commonly raised, the periphery more acute and the walls lack the shining appearance of those of our *R. beccarii* individuals.

Distribution. KRUIT reported that this species is mainly present in shallow sandy deposits at depths of less than 13 m. He suggested that the specimens might be subrecent (1954).

Originally *R. sarmientoi* was described from the Miocene Tubara beds of northern Columbia.

In our material the distribution of *R. sarmientoi* roughly coincides with that of *R. beccarii* vars. There is some difference in relative abundance of both species, *R. sarmientoi* being more frequent in the deposits off the

Essequibo than in those off the Orinoco. In this connection KRUIT's suggestion that the species in the Gulf of Paria might be subrecent needs serious consideration, the more so since it is identified with a Miocene Columbian species. However, another explanation of the different relative distribution might be a greater preference of *R. sarmientoi* for more sandy sediments.

If depth is considered to be the main controlling factor for the dominance of both *Rotalia* species, the close connection of their distribution on the Orinoco shelf with the 20 fathom line would cause no difficulty. If salinity fluctuations are the main factor, which is considered more likely (see *R. rolshauseni*), then the outer part of the Essequibo area may contain numerous subrecent specimens of both species. If only *R. sarmientoi* would be subrecent, this would less easily explain its greater frequency in the entire Essequibo area, though resedimentation seems to be fairly common. In this concept the high frequency of *R. sarmientoi* east of Trinidad has no explanation. The latter occurrence would better fit in with a concept of dominance by more sandy sediment. Different information is found in data from two of the cores near Trinidad (see KOLDEWIJN). At EC 1197 a deep sample (225 cm) with a fauna different from the surface one, hence at least subrecent, shows more *R. sarmientoi* than *R. beccarii*. This might support the assumed fossil character of the former species in surface material. At DR 1106 (185 cm) in clay with a fauna strongly deviating from the surface one, 17 % of *R. beccarii* was found and no *R. sarmientoi*. This again favours the assumption of dependence on sediment type.

Altogether the latter theory seems somewhat more plausible, but the influence of different subrecent relations of both species cannot be excluded.

Rotalia translucens PHLEGER and PARKER

Rotalia translucens PHLEGER and PARKER, 1951, Geol. Soc. Am., Mem. 46, pt. 2, p. 24, pl. 12, fig. 11, 12.

Distribution. Rare occurrences only, mainly along the continental slope, but as shallow as 62 metres.

Saracenaria italica DEFRANCE

Cristellaria italica (DEFRANCE), H. B. BRADY, 1884, Rep. Voy. Challenger, Zoology, vol. 9, p. 544, pl. 68, fig. 17, 18, 20–23.

Remarks. Some indistinct specimens which may have some relation with our *Marginulina-Vaginulina* group.

Distribution. Single specimens at a small number of stations, mainly along the shelf border, but also as shallow as 68 metres.

Seabrookia earlandi WRIGHT

Textfig. 4

Seabrookia earlandi WRIGHT, 1891, Proc. Roy. Irish Ac., ser. 3, vol. 1, p. 477, pl. 20, fig. 6, 7; CUSHMAN, 1924, U.S. Nat. Mus., Bull. 104, pt. 5, p. 5, pl. 1, fig.

14–16; PARKER, 1954, Bull. Mus. Comp. Zool. Harv. Coll., vol. 111, no. 10, p. 538, pl. 11, fig. 13.

Distribution. PARKER reported this species with very low frequencies at all depths. PHLEGER (1956) found eight living specimens between 63 and 81 metres.

In our material scattered from 10 to 200 metres, but there is a remarkable area with higher frequencies (up to 4 %) between 65 and 90 metres on the outer shelf off the Orinoco, in pelitic sediments with calcarenitic and sandy admixtures.

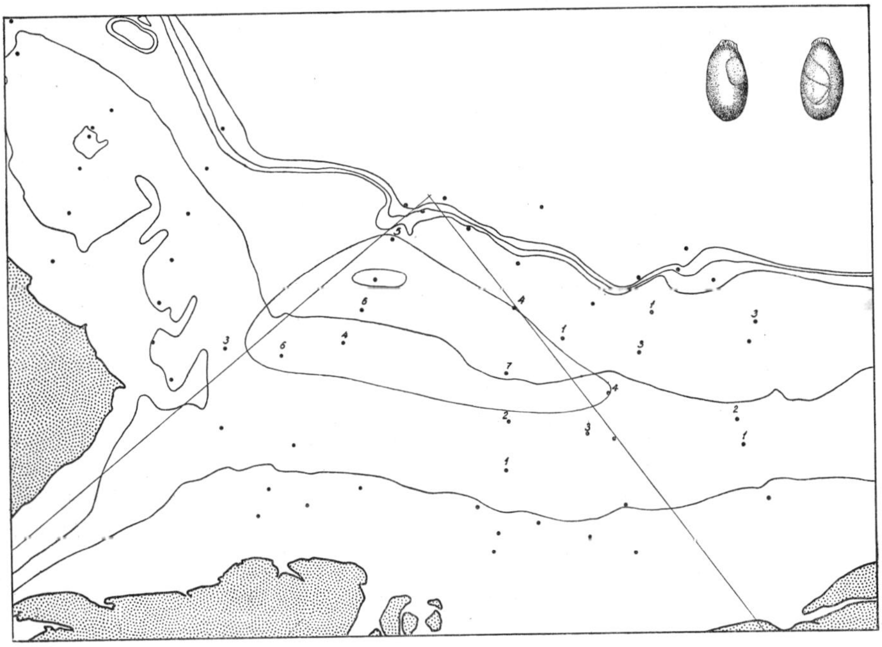

Fig 4.

Distribution of *Seabrookia earlandi* in part of the investigated area. Figures ×65. Station number DH 1040.

Sigmoilina tenuis (CZJZEK)

Quinqueloculina tenuis CZJZEK, 1848, Haidinger's Nat. Abh., vol. 2, p. 149, pl. 13, fig. 31–34.

Sigmoilina tenuis (CZJZEK), PHLEGER and PARKER, 1951, Geol. Soc. Am., Mem. 46, pt. 2, p. 8, pl. 4, fig. 7; PARKER, 1954, Bull. Mus. Comp. Zool. Harv. Coll., vol. 111, no. 10, p. 499, pl. 4, fig. 19.

Distribution. At 38 m. and deeper. Frequencies of 2 % and slightly more are found deeper than 75 m., mostly along the border of the shelf.

Siphonina reticulata (CZJZEK)

Pl. 3, fig. 11

Rotalina reticulata CZJZEK, 1848, Haidinger's Nat. Abh., vol. 2, p. 145, pl. 13, fig. 7, 8.

Siphonina reticulata (CZJZEK), MARKS, 1951, Contr. Cushm. Found. Foram. Res., vol. 2, p. 65, pl. 8, fig. 8; KEY, 1954, Verh. K.N.A.W., ser. 1, vol. 20, no. 5, p. 215, pl. 3, fig. 1.

Remarks. If the wide size range is taken into account, many of the stated differences between described recent *Siphonina* species are rendered illusionary. Typical *S. bradyana* CUSHMAN (1927, Proc. U.S. Nat. Mus., vol. 72, art. 20, p. 11, pl. 1, fig. 4) are scarce in our material, but variation in the fimbriation and wideness of the keel are present. Some specimens show the fine tubules that do not reach the edge of the keel, others have a fully developed fimbriation. Transitions between these two types are also present. As a consequence, the specimens figured by PHLEGER and PARKER (1951, pl. 12, fig. 13, 14) are certainly within the range of variation of our *S. reticulata*. Large specimens may have a more or less rounded periphery, a thicker wall and more evenly scattered large pores; they are identical with *S. pulchra* CUSHMAN (1919, Carnegie Inst. Wash., Publ. 291, p. 42, pl. 14, fig. 17). We united in our material all these forms in *S. reticulata.*

Distribution. In the Gulf of Mexico PHLEGER and PARKER (1951) found *Siphonina* most common between 30 and 225 m. PARKER (1954) reported *S. bradyana* and *S. pulchra* most frequent between 100 and 300 m. PHLEGER (1956) found *S. pulchra* at 42 m. and deeper, more constant at 68 m. and deeper. Frequencies never more than 2 %.

In our material scattered occurrences at all depths. Greater frequencies, up to 4 %, at 35 m. and deeper.

Siphotextularia cf. **S. affinis** (FORNASINI)

Cf. *Sagrina affinis* FORNASINI, 1883, Boll. Soc. Geol. Ital., vol. 2, p. 189, pl. 2, fig. 10.
Karreriella affinis (FORNASINI), CUSHMAN, 1937, Cushm. Lab. Foram. Res., Spec. Publ. no. 8, p. 134, pl. 16, fig. 2, 3.

Remarks. We found a small number of different *Siphotextularia* specimens in nearly as many samples. Some are identical with those figured by CUSHMAN from the Pliocene of Sicily.

Distribution. Scattered occurrences between about 15 and about 200 metres.

Sphaeroidina bulloides D'ORBIGNY

Pl. 3, fig. 12

Sphaeroidina bulloides D'ORBIGNY, 1826, Ann. Sci. Nat., vol. 7, p. 267, mod. 65; CUSHMAN and TODD, 1949, Contr. Cushm. Lab. Foram. Res., vol. 25, p. 13, pl. 3, fig. 8–11; PARKER, 1954, Bull. Mus. Comp. Zool. Harv. Coll., vol. 111, no. 10, p. 539, pl. 11, fig. 18.

Distribution. According to PARKER the species occurs in the Gulf of Mexico deeper than 100 m.; between 180 and 1000 m. frequencies are highest, up to 9 % at 314 m.

In our material the species follows the continental slope of the Orinoco shelf. North of Trinidad-Paria the occurrences are at 100 m. and deeper. Highest frequencies at 135 m. and deeper (up to 7 %), mostly in pelite and pelitic sediment.

Spirillina vivipara EHRENBERG

Spirillina vivipara EHRENBERG, 1841, Abh. k. Ak. Wiss. Berlin, p. 422, pl. 3, sec. 7, fig. 41; H. B. BRADY, 1884, Rep. Voy. Challenger, Zoology, vol. 9, p. 630, pl. 85, fig. 1–4; CUSHMAN, 1931, U.S. Nat. Mus., Bull. 104, pt. 8, p. 3, pl. 1, fig. 1–4; PHLEGER and PARKER, 1951, Geol. Soc. Am., Mem. 46, pt. 2, p. 25, pl. 13, fig. 3, 4.

Remarks. Our few *Spirillina* specimens show considerable variation, probably ranging beyond the limits of *S. vivipara*. Also conical individuals have been included.

Distribution. Some 15 scarce occurrences, mostly between 45 and 95 metres.

Spiroloculina depressa D'ORBIGNY

Spiroloculina depressa D'ORBIGNY, 1826, Ann. Sci. Nat., vol. 7, p. 298, mod. 92; CUSHMAN and TODD, 1944, Cushm. Lab. Foram. Res., Spec. Publ. no. 11, p. 28, pl. 1, fig. 1, 6, pl. 5, fig. 1–9.

Remarks. Specimens of other *Spiroloculina* species may have been incorporated.

Distribution. Frequencies less than 1 %, scattered at some 15 stations between 30 and 95 metres; one occurrence at 153 m.

Spiroplectammina floridana (CUSHMAN)

Map 32

Textularia floridana CUSHMAN, 1922, Carnegie Inst. Wash., Publ. no. 311, p. 24, pl. 1, fig. 7; CUSHMAN, 1922, U.S. Nat. Mus., Bull. 104, pt. 3, p. 18, pl. 2, fig. 11, 12.

Spiroplectammina floridana (CUSHMAN), PHLEGER and PARKER, 1951, Geol. Soc. Am., Mem. 46, pt. 2, p. 4, pl. 1, fig. 25, 26; PARKER, 1954, Bull. Mus. Comp. Zool. Harv. Coll., vol. 111, no. 10, p. 490, pl. 2, fig. 9.

Distribution. In CUSHMAN's Atlantic Ocean report most occurrences are from 40 to 140 m., rare occurrences down to 2200 m. PHLEGER and PARKER report the species from calcareous areas.

In our material there is a distinct band with maximal frequencies (up to 5 %) on the Orinoco shelf between 60 and 110 m. In the Trinidad-Paria region some occurrences are found between 45 and 65 m., all in more or less calcarenitic sand. A few other high frequencies are found along the continental slope.

Stetsonia minuta PARKER

Stetsonia minuta PARKER, 1954, Bull. Mus. Comp. Zool. Harv. Coll., vol. 111, no. 10, p. 534, pl. 10, fig. 27–29.

Distribution. PHLEGER (1956) found a range of 16–101 m., and living specimens at 16, 23 and 66 metres.

In our material with frequencies up to 5% shallower than 60 m.; but near Tobago a frequency of 3% was found as deep as 180 m. The other occurrences in the Trinidad-Paria region are all somewhat deeper than they are on the Orinoco shelf.

Stomatorbina concentrica (PARKER and JONES)

Pulvinulina concentrica PARKER and JONES MS, 1864, in H. B. BRADY, Trans. Linn. Soc. Zool., vol. 24, p. 470, pl. 48, fig. 14; H. B. BRADY, 1884, Rep. Voy. Challenger, Zoology, vol. 9, p. 686, pl. 105, fig. 1.

Stomatorbina concentrica (PARKER and JONES), PHLEGER and PARKER, 1951, Geol. Soc. Am., Mem. 46, pt. 2, p. 22, pl. 12, fig. 2.

Distribution. Some 10 occurrences of one or two specimens between 52 and more than 183 m., only in calcarenitic sediments.

Textularia candeiana D'ORBIGNY

Map 33

Textularia candeiana D'ORBIGNY, 1839, in Ramon de la Sagra, Hist. Fis. Pol. Nat. Cuba, Foraminifères, p. 143, pl. 1, fig. 25–27; CUSHMAN, 1922, Carnegie Inst. Wash., Publ. 311, p. 23, pl. 2, fig. 2; CUSHMAN, 1922, U.S. Nat. Mus., Bull. 104, pt. 3, p. 8, pl. 1, fig. 1–3; BANDY, 1954, U.S. Geol. Survey, Prof. Paper 254–F, p. 139, pl. 29, fig. 2; PARKER, 1954, Bull. Mus. Comp. Zool. Harv. Coll., vol. 111, no. 10, p. 490, pl. 2, fig. 16, 17.

Remarks. Our *Textularia* specimens show considerable variation in size and shape of the test. The most distinct type among the larger individuals is that of *T. candeiana*. CUSHMAN called their shape club-like. The early part of the test is rather smooth, compressed and narrow. The later chambers rapidly increase in all three dimensions, while the wall commonly becomes more coarsely arenaceous. Possibly some of the specimens have a somewhat coiled initial part.

The main difficulty is in the determination of small specimens, which did not yet reach the typical *candeiana* stage. Among them the types of *T. conica* D'ORBIGNY and even sometimes *T. mayori* CUSHMAN may be recognized. Especially the former is also found among specimens that are comparable in size to smaller typical *T. candeiana* specimens.

Distribution. Records in the literature are vague: PARKER (1948), rare but typical between 90 and 300 m.; PARKER (1954), consistent to a depth of 345 m., frequencies less than 5 %; BANDY (1954), greatest abundance on prominences with a marked increase in median grain size.

For our material, see under *T. conica*.

Textularia conica D'ORBIGNY

Map 33

Textularia conica D'ORBIGNY, 1839, in Ramon de la Sagra, Hist. Fis. Pol. Nat. Cuba, Foraminifères, p. 143, pl. 1, fig. 19, 20; H. B. BRADY, 1884, Rep. Voy. Challenger, Zoology, vol. 9, p. 365, pl. 43, fig. 13, 14, pl. 113, fig. 1; PHLEGER and PARKER, 1951, Geol. Soc. Am., Mem. 46, pt. 2, p. 5, pl. 1, fig. 27; PARKER, 1954, Bull. Mus. Comp. Zool. Harv. Coll., vol. 111, no. 10, p. 490, pl. 2, fig. 13.

Remarks. Typical specimens differ from *T. candeiana* by the lesser relative height of the later chambers. Mostly they are of smaller size, but otherwise both types grade into one another.

Occasional specimens resemble *T. pseudogramen* CHAPMAN and PARR (see BROOKS ELLIS, and BRADY, 1884, pl. 43, fig. 9) and *T. mayori* CUSHMAN, as figured by PHLEGER and PARKER (1951, pl. 2, fig. 1–5), though distinct fistules are lacking.

Distribution. PARKER (1954) reported *T. conica* down to a depth of 280 m., with frequencies of 2 % and less.

In our material the combined *T. candeiana* and *T. conica* show a distinct preference for the coarser calcarenitic sediments. East of Trinidad and north of the Essequibo the high frequencies (up to 15 %) are shallower than 70 m. On the remainder of the Orinoco shelf they are deeper than 70 m., occasionally reaching the continental slope. Except for some samples in shallow water near the Trinidad coast, *T. conica* is always the dominant form.

Textularia foliacea HERON-ALLEN and EARLAND var. **occidentalis** CUSHMAN

Pl. 3, fig. 13, 14

Textularia foliacea HERON-ALLEN and EARLAND var. *occidentalis* CUSHMAN, 1922, U.S. Nat. Mus., Bull. 104, pt. 3, p. 16, pl. 2, fig. 13; PHLEGER and PARKER, 1951, Geol. Soc. Am., Mem. 46, pt. 2, p. 5, pl. 1, fig. 28, 29; PARKER, 1954, Bull. Mus. Comp. Zool. Harv. Coll., vol. 111, no. 10, p. 491, pl. 2, fig. 10.

Distribution. Low frequencies (up to 3 %) at 55 m. and deeper.

Textularia mexicana CUSHMAN

Pl. 3, fig. 15

Textularia mexicana CUSHMAN, 1922, U.S. Nat. Mus., Bull. 104, pt. 3, p. 17, pl. 2, fig. 9; PHLEGER and PARKER, 1951, Geol. Soc. Am., Mem. 46, pt. 2, p. 5, pl. 2, fig. 6, 7.

Distribution. Some occurrences of 1 % or less, usually deeper than 140 m., but one at a depth of 70 m.

Textulariella barrettii (JONES and PARKER)

Pl. 4, fig. 1

Textularia barrettii JONES and PARKER, 1863, Rep. Brit. Ass., Newcastle meeting, p. 80, 105; H. B. BRADY, 1884, Rep. Voy. Challenger, Zoology, vol. 9, p. 367, pl. 44, fig. 6–8; CUSHMAN, 1922, U.S. Nat. Mus., Bull. 104, pt. 3, p. 20, pl. 3, fig. 3–6.

Distribution. PARKER (1954) possibly included this species in *Textulariella* spp., which group she found to a depth of 180 m., with one exception at 320 m.

In our material with an upper depth limit of 65 m., mostly on the outer part of the shelf. Frequencies are variable, usually low; the maximum is 10 % at DZ 1173.

Trifarina bradyi CUSHMAN

Rhabdogonium tricarinatum, H. B. BRADY (not D'ORBIGNY), 1884, Rep. Voy. Challenger, Zoology, vol. 9, p. 525, pl. 67, fig. 1–3.

Trifarina bradyi CUSHMAN, 1923, U.S. Nat. Mus., Bull. 104, pt. 4, p. 99, pl. 22, fig. 3–9; PHLEGER and PARKER, 1951, Geol. Soc. Am., Mem. 46, pt. 2, p. 18, pl. 8, fig. 10, 11; PARKER, 1954, Bull. Mus. Comp. Zool. Harv. Coll., vol. 111, no. 10, p. 522, pl. 8, fig. 9.

Remarks. Only a few of our specimens distinctly belong to *T. bradyi*. The number of uniserial chambers in most of our individuals is distinctly smaller than in those figured by BRADY, CUSHMAN, and PHLEGER and PARKER. Correspondingly they are mostly less elongate and often with more strongly downward curved angles of the chambers. Especially the smaller individuals are closer to *T. reussi* (CUSHMAN) (see BRADY, pl. 67) than to *T. bradyi*. They are often more or less *Angulogerina*-like. Possibly we are dealing with mainly immature individuals of *T. bradyi*, though partly of considerable size. More material would be needed for a correct determination, since we have only one or a few specimens of the various stations.

Distribution. Scattered occurrences, mostly between 75 and 200 metres, always with low frequencies.

Triloculina tricarinata D'ORBIGNY

Triloculina tricarinata D'ORBIGNY, 1826, Ann. Sci. Nat., ser. 1, vol. 7, p. 299, no. 7, mod. 94; CUSHMAN, 1929, U.S. Nat. Mus., Bull. 104, pt. 6, p. 56, pl. 13, fig. 3.

Distribution. Some 10 occurrences scattered between 40 and 90 m., always less than 1 %.

Triloculina trigonula (LAMARCK)

Miliolites trigonula LAMARCK, 1804, Ann. Mus., vol. 5, p. 351, no. 3.

Triloculina trigonula (LAMARCK), CUSHMAN, 1929, U.S. Nat. Mus., Bull. 104, pt. 6, p. 56, pl. 12, fig. 10, 11, pl. 13, fig. 1, 2.

Distribution. KRUIT (1954) found scattered specimens in the Serpents Mouth and on the Soldado platform. He concluded to a restriction to saline environments at less than 35 m.

In our material with low frequencies at some 25 stations between 20 and 110 m., with one exception at 183 m.

Trochammina advena CUSHMAN

Trochammina advena CUSHMAN, 1922, Carnegie Inst. Wash., Publ. no. 311, p. 20, pl. 1, fig. 2–4; PHLEGER and PARKER, 1951, Geol. Soc. Am., Mem. 46, pt. 2, p. 9, pl. 4, fig. 15; PARKER, 1954, Bull. Mus. Comp. Zool. Harv. Coll., vol. 111, no. 10, p. 502, pl. 5, fig. 5, 6.

Distribution. Single specimens in a few samples down to about 75 metres.

Uvigerina flinti CUSHMAN

Pl. 4, fig. 2

Uvigerina flintii CUSHMAN, 1923, U.S. Nat. Mus., Bull. 104, pt. 4, p. 165, pl. 42, fig. 13; PHLEGER and PARKER, 1951, Geol. Soc. Am., Mem. 46, pt. 2, p. 18, pl. 8, fig. 15, 16.

Distribution. In the Gulf of Mexico PHLEGER and PARKER (1951) found the species restricted to the range 100–200 m., but PARKER (1954) found it from 220 to 1000 m., with frequencies up to 13 %.

In our material with one exception always at 125 m. and deeper. In many samples along the shelf border; frequencies usually low, but ranging up to 8 %.

Uvigerina peregrina CUSHMAN

Map 34

Uvigerina peregrina CUSHMAN, 1923, U.S. Nat. Mus., Bull. 104, pt. 4, p. 166, pl. 42, fig. 7–10.

Remarks. Many of our samples yielded a rich material of costate Uvigerinae. Although distinct differences may be noted between individual specimens as well as between groups of individuals from different samples, we failed to separate really different species on morphological grounds.

All individuals are characterized by moderately inflated chambers arranged in a triserial series and more or less tending to become uniserial. The wall of the chambers is ornamented with non-continuous costae; costae and wall being finely perforate with occasional larger pores, or pitted or covered with minute granules. The costae are plate-like, more or less high, more or less broken up in the later part of the test, regularly spaced, the intervals between them generally being some times wider than the width of the costae. The aperture is placed at the end of a distinct neck, in well-preserved specimens provided with a slight flaring lip.

Features, generally used in separating the various species of this group show no constancy in our material. The length varies from one to four times the breadth. In short forms the chambers are less remote, better retaining the triserial arrangement; the greatest breadth is generally situated above the middle. In more elongate forms the triserial arrangement of the chambers is more drawn out, tending to irregularly biserial or uniserial; the greatest breadth is mostly near the middle of the test, which has more or less parallel sides in this part. In part of the specimens the later chambers become so remote that the breadth of the later portion of the test is less than in the more close-coiled earlier parts.

Features that show some correlation with the geographic location of the stations are spinosity of the wall and size of the test, and to a lesser extent the relation between length and breadth of the test.

With the exception of some extremely big specimens (in DH 1037, referred with some doubt to this species), decrease of size of the test seems to depend on life conditions, since this decrease of individual size roughly corresponds with decrease in relative frequency of the species. Variation of relative length is less clear (elongate forms assignable to *U. peregrina* var. *bradyana* CUSHMAN), but spinosity also seems to increase with decrease in relative frequency of the species. It is even a better distinctive character than size.

As specific name that of *U. peregrina* has been given to the entire group of specimens. Comparison of these specimens with those of *U. hispido-costata* CUSHMAN and TODD from the Miocene of Aruba failed to yield very distinct differences. There are occasional differences in average features, such as the place of the greatest breadth of the test, which is more towards the apertural end in the Miocene specimens. The granular surface of *U. peregrina* is apparently no constant character, both the Miocene and the recent specimens showing the same variation in this feature. Large suits of specimens from various deposits are needed to decide whether both species are conspecific or not.

For separation of the variants in a rough way, *U. peregrina peregrina* is taken for the larger, more stoutly built specimens with costae over the entire test, only occasionally broken up into spines on the later chambers. Smaller, more spinose individuals are generally referred to as *U. peregrina parvula*. Of course the boundary is arbitrary.

Distribution. See *U. peregrina parvula.*

Uvigerina peregrina CUSHMAN var. **parvula** CUSHMAN

Map 35

Uvigerina peregrina CUSHMAN var. *parvula* CUSHMAN, 1923, U.S. Nat. Mus., Bull. 104, pt. 4, p. 168, pl. 42, fig. 11; PHLEGER and PARKER, 1951, Geol. Soc. Am., Mem. 46, pt. 2, p. 18, pl. 8, fig. 27–30.

Uvigerina parvula CUSHMAN, PARKER, 1954, Bull. Mus. Comp. Zool. Harv. Coll., vol. 111, no. 10, p. 521, pl. 8, fig. 6.

Remarks. This variety is usually smaller than the typical, with lower costae that are mostly strongly broken up on a variable portion of the test. The later chambers are often only spinose to finely hispid; they are commonly more remote than in *U. peregrina peregrina.*

Our interpretation of this variety is identical with that given by PHLEGER and PARKER, which does not exactly correspond to CUSHMAN's diagnosis.

Distribution. *U. peregrina peregrina* was reported by CUSHMAN (1923) as a cold water species off the eastern coast of the United States. PHLEGER and PARKER (1951) found the species characteristic of depths between 100 and 1850 m.; at a large number of stations deeper than 500 m., frequencies were mostly between 5 and 20 %. PARKER (1954) found relatively high frequencies between 200 and 1600 m., up to 25 %. WALTON reported living specimens from Todos Santos Bay, California, between 36 and 1100 m. KRUIT (1954) found the species abundant in the area of Bocas del Dragon in marine muds, commonly deeper than 20 m., possibly tolerating some salinity decrease.

U. peregrina parvula was found 10 to 35 % by PHLEGER and PARKER (1951) between 45 and 135 m., by PARKER (1954) with highest frequencies shallower than 290 m. PHLEGER (1956) reported living specimens from 58–110 metres, frequencies of dead ones 10–25 %.

U. peregrina parvula is generally reported from shallower water than *U. peregrina peregrina.*

In our material *U. peregrina peregrina* reaches its maximal frequencies in pelite and highly pelitic sediments. Evidently it avoids notable salinity fluctuations. An area of very high percentages is found in the pelite, north of Trinidad-Paria (up to 81 % at 80 m.) from about 40 m. to about 150 m. depth. A strip of frequencies up to 15 % is found in the pelite area off the Orinoco mouth, between 40 and 70 m., clearly avoiding the *Rotalia* zone. At several places along the continental slope high frequencies reappear (up to 32 %), again in very pelitic sediments. The occurrence of nearly 80 % at DZ 1177 is not recent, which may be seen from the preservation of the specimens.

U. peregrina parvula is most frequent (up to 15 %) in the more sandy zone of the Orinoco shelf in between the inner and outer areas of *U. peregrina peregrina.* This band continues north of Trinidad-Paria, but here it mainly overlaps the area of the other variant.

In a wide area off the Essequibo both variants are conspicuously absent. East of Trinidad they are also very rare.

Especially if relative proportions of the variants are regarded we find a strong dominance of *U. peregrina parvula* in the more sandy deposits around the 75 m. line on the Orinoco shelf, and in similar areas near Tobago and on approaching Cumberland Bank. In the pelite areas *U. peregrina peregrina* is dominant. Evidently, the muddy character is the main factor in the distribution of the latter. This distribution seems to be fairly independent of depth, though areas with distinct salinity fluctuations are probably avoided. It is considered likely that the individuals are "mud eaters". As soon as conditions become worse because of the admixture of sandy material, the small variant *U. peregrina parvula* is developed, while numbers of individuals diminish. In too sandy deposits this variant is also lacking.

Uvigerina peregrina CUSHMAN var.

Remarks. Variety differing from the typical in the much larger size of the test. It occurs only in DH 1037 (1 %). It seems to be identical with the big specimens identified by PHLEGER and PARKER as *U. hispido-costata* CUSHMAN and TODD (1951, p. 18, pl. 8, fig. 20, 21, 23).

Uvigerina proboscidea SCHWAGER

Map 36

Uvigerina proboscidea SCHWAGER, 1866, Novara Exp., Geol. Theil., vol. 2, pt. 2, p. 250, pl. 7, fig. 96; CUSHMAN and TODD, 1941, Contr. Cushm. Lab. Foram. Res., vol. 17, p. 73, pl. 17, fig. 9, pl. 19, fig. 3–9; DROOGER, 1953, Contr. Cushm. Found. Foram. Res., vol. 4, p. 136, pl. 21, fig. 28.

Uvigerina auberiana D'ORBIGNY var. *laevis* GOËS, PHLEGER and PARKER, 1951, Geol. Soc. Am., Mem. 46, pt. 2, p. 18, pl. 8, fig. 12–14.

Remarks. There is considerable variation in the relative length of the test and in the coarseness of the granules and hispidation. The hispid character of the wall and the lack of any striation serve to distinguish this species from spinose variants of *U. peregrina*, though occasional very finely striated specimens of the latter's *parvula* variety are close to our *U. proboscidea*. From the Gulf of Mexico the species is usually recorded under other specific names, but in our opinion there is no necessity to subdivide our group of specimens.

Distribution. PHLEGER and PARKER (1951) found a depth range from 30 to 300 m., most abundant between 100 and 210 m., frequencies up to over 25 %. PARKER (1954) found highest frequencies between 160 and 275 m., PHLEGER (1956) living and dead specimens at 85 m. and deeper.

In our material the species occurs at 50 m. and deeper. Frequencies of 2 % and more are deeper than 90 m. Most of them occur at scattered places along the border of the shelf in its Orinoco part. Others (up to 17 % at 145 m.) are found in the area of mixed pelite and calcarenite north of Trinidad deeper than 120 metres.

Vaginulina planata (PHLEGER and PARKER)

Pl. 4, fig. 3

Marginulina planata PHLEGER and PARKER, 1951, Geol. Soc. Am., Mem. 46, pt. 2, p. 9, pl. 4, fig. 21, 22, pl. 5, fig. 1–3.

Distribution. PHLEGER and PARKER found a depth range of 30 to 90 m., and frequencies up to 3 %.

In our material the depth range is from 64 to over 183 m. Frequencies up to 5 %, at DG 1028 (80 m.). Most occurrences on the outer shelf in coarse sediments.

Valvulineria humilis (BRADY)

Truncatulina humilis H. B. BRADY, 1884, Rep. Voy. Challenger, Zoology, vol. 9, p. 665, pl. 94, fig. 7.

Valvulineria humilis (BRADY), PHLEGER and PARKER, 1951, Geol. Soc. Am., Mem. 46, pt. 2, p. 25, pl. 13, fig. 9, 10.

Remarks. Our specimens are in good agreement with the figures and remarks given by BRADY and by PHLEGER and PARKER. Often they are hardly separable from small five-chambered *Globigerina* specimens.

Distribution. At all depths, but mostly in deeper water. The maximum frequency is 10 %, but usually frequencies are 1 % or less.

Valvulineria minuta PARKER

Valvulineria minuta PARKER, 1954, Bull. Mus. Comp. Zool. Harv. Coll., vol. 111, no. 10, p. 527, pl. 9, fig. 4–6.

Distribution. Scattered specimens at 68 m. and deeper.

Valvulineria sp.

Remarks. A number of small specimens that resemble *Anomalinoides mexicana* PARKER (1954, Bull. Mus. Comp. Zool. Harv. Coll., vol. 111, no. 10, p. 539, pl. 11, fig. 21–23), but dorsally they are slightly evolute.

Distribution. Some 15 occurrences at variable depth, shallowest at 37 m.

Virgulina antilleana nom. nov.

Pl. 4, fig. 4

Bolivina pulchella (D'ORBIGNY) var. *primitiva* CUSHMAN, 1930, Florida State Dept. Geol. Survey, Bull. 4, p. 47, pl. 8, fig. 12; CUSHMAN, 1937, Cushm. Lab. Foram. Res., Spec. Publ. no. 9, p. 90, pl. 12, fig. 6; PHLEGER and PARKER, 1951, Geol. Soc. Am., Mem. 46, pt. 2, p. 14, pl. 7, fig. 3; PARKER, 1954, Bull. Mus. Comp. Zool. Harv. Coll., vol. 111, no. 10, p. 516, pl. 7, fig. 36.

Remarks. The triserial portion of our specimens is variously developed, sometimes being nearly absent. The test is distinctly twisted, however. Occasional specimens have the broader appearance of *Bolivina pulchella*, as figured by CUSHMAN (1937, p. 151, pl. 15, fig. 10, 11).

Because of the triserial part of the test, the species had better be removed from the genus *Bolivina* and placed in *Virgulina*, which of course does not exclude its ancestry of *B. pulchella*. Since the name *Virgulina primitiva* has been preoccupied by CUSHMAN (1936, Cushm. Lab. Foram. Res., Spec. Publ. no. 6, p. 46, pl. 7, fig. 1), a new name had to be given.

Distribution. According to PHLEGER and PARKER (1951), restricted to water less than about 100 m. depth, most abundant less than 80 m. PARKER (1954) noted occurrences at all depths.

In our material scattered distribution and low frequencies (occasionally up to 4 %), mostly shallower than 80 m., but some occurrences near the continental slope were found as well.

Virgulina complanata EGGER

Pl. 4, fig. 5

Virgulina schreibersiana CZJZEK var. *complanata* EGGER, 1883, Abh. k. bay. Ak. Wiss. München, vol. 18, p. 292, pl. 8, fig. 91, 92.

Virgulina complanata EGGER, CUSHMAN, 1937, Cushm. Lab. Foram. Res., Spec. Publ. no. 9, p. 26, pl. 4, fig. 13–17; PHLEGER and PARKER, 1951, Geol. Soc. Am., Mem. 46, pt. 2, p. 19, pl. 9, fig. 1–3; PARKER, 1954, Bull. Mus. Comp. Zool. Harv. Coll., vol. 111, no. 10, p. 512, pl. 7, fig. 6.

Distribution. Fairly widespread, with low frequencies (up to 2 %), but always at depths less than 100 m.

Virgulina pontoni CUSHMAN

Map 37

Virgulina pontoni CUSHMAN, 1932, Contr. Cushm. Lab. Foram. Res., vol. 8, p. 17, pl. 8, fig. 7; CUSHMAN, 1937, Cushm. Lab. Foram. Res., Spec. Publ. no. 9, p. 19, pl. 2, fig. 26–28; PHLEGER and PARKER, 1951, Geol. Soc. Am., Mem. 46, pt. 2, p. 19, pl. 9, fig. 9, 10; KEY, 1954, Verh. K.N.A.W., ser. 1, vol. 20, no. 5, p. 212, pl. 2, fig. 3; PARKER, 1954, Bull. Mus. Comp. Zool. Harv. Coll., vol. 111, no. 10, p. 513, pl. 7, fig. 9.

Distribution. PHLEGER and PARKER (1951) found frequencies up to 35 % at less than 125 m.; PARKER (1954) as high as 4 % shallower than 105 m.; PHLEGER (1956) 1–15 % between 16 and 110 m., but greatest frequencies between 23 and 75 m. In the Gulf of Paria the species was found to be common to abundant in muds less than 35 m. deep.

In our material low frequencies occur all over the investigated area, but a very distinct axis with frequencies up to 60 % was found at approximately 35–40 m. in the pelite area off the Orinoco.

Virgulina spinicostata PHLEGER and PARKER

Pl. 4, fig. 6

Virgulina spinicostata PHLEGER and PARKER, 1951, Geol. Soc. Am., Mem. 46, pt. 2, p. 19, pl. 9, fig. 11–14.

Remarks. *V. spinicostata* somewhat resembles *Bolivina pulchella* (D'ORBIGNY) and *B. pulchella primitiva* CUSHMAN, as described and figured by CUSHMAN in his monograph of the Virgulininae (1937, Cushm. Lab. Foram. Res., Spec. Publ. no. 9, p. 151, pl. 15, fig. 9–11, and p. 90, pl. 12, fig. 6 respectively). The test is more rounded, the chambers are relatively higher and the costae less prominent.

Distribution. In the Gulf of Mexico according to PHLEGER and PARKER (1951) up to 2 % down to 80 m. PHLEGER (1956) gives a range from 11 to 99 m., average frequencies of 2 to 4 % between 40 and 80 m.

In our material scattered distribution, mostly shallower than 100 m., frequencies up to 3 %.

Planktonic species

Globigerinidae

Maps 38, 39

The Globigerinidae of our samples belong to a great number of types. Several of these types clearly belong together as variants of a single species. These relations are most distinct among the bigger, well-developed individuals. Middle-sized specimens can generally be referred without much difficulty to one of the adult types, but especially among the smaller individuals there are various types, for which specific names are difficult to be found in the literature. These small-sized individuals and types may be of local importance in many instances.

Among the larger individuals the majority belong to the genus *Globigerinoides*. The adult types are clearly interrelated.

Especially among the middle-sized individuals **Globigerinoides cyclostoma** (GALLOWAY and WISSLER) (pl. 5, fig. 7) is the most frequent. This is a rather compact form with three chambers in the final coil, hardly elevated spiral, later chambers not very much increasing in size and often slightly flattened, with circular to semicircular last aperture, directly opposite the suture between the previous two chambers; earlier apertures very small; coiling is nearly in a single plane. This species is but slightly different from *G. subquadrata* BRONNIMANN, which seems to have a more compact form and a somewhat smaller average aperture.

Globigerinoides triloba (REUSS) (pl. 5, fig. 8), also very frequent, usually reaches greater size. It differs from *G. cyclostoma*, with which it intergrades, in the more rapid increase in size of the chambers so that the final chamber is larger than the entire visible earlier part of the test. The aperture is usually lower, more elongate and mainly lying to one side of the suture between the two previous chambers.

Globigerinoides sacculifera (BRADY) (pl. 5, fig. 9) is another frequent variant of great size, in which the final chamber is of variable size, and elongated mostly in peripheral direction. Such a chamber, occasionally more than one, is superimposed on earlier tests of both previous types. It may have been added obliquely to the plane of coiling, mostly ventrally.

In **Globigerinoides rubra** (D'ORBIGNY) (pl. 5, fig. 10) several of the later chambers are moderately or slightly increasing in size, and added in a ventrally oblique direction so that the test becomes more highly trochoid. Apertures are usually large and the later chambers are often distinctly flattened. These specimens, which are rather frequent among the middle- and large-sized individuals, show distinct intergradation with the other types, mainly with that of *G. cyclostoma*. Except for the reddish colour, the original *G. rubra* seems to be very close to *G. triloba*. Therefore it might be better to call our variant *G. pyramidalis* (VAN DEN BROECK), but for unknown reasons most authors give it the name established by D'ORBIGNY.

"**Abnormal" Globigerinoides** (pl. 5, fig. 11) are rather frequent individuals of various sizes, in which the final chamber(s) is exceedingly small, mostly thin-walled and with distinct, proportionally large apertures. Such chambers are mostly added to earlier parts that resemble the *cyclostoma* type. Also some wild-growing individuals have been included here.

Globigerinoides conglobata (BRADY) (pl. 5, fig. 12) in its typical globular form is rare. The individuals are thick-walled. Among the smaller, less globular and less thick-walled specimens there is imperceptible gradation to *G. cyclostoma*, as may also be concluded from BRADY's figures.

Sphaeroidinella dehiscens (PARKER and JONES) (pl. 5, fig. 13), represented only by very few, distinct, large individuals, is another thick-walled species which may belong to this group. It has the shape and chamber arrangement of *G. triloba*, but all earlier features are obliterated by the excessive shell-growth.

Small individuals of *Globigerinoides* generally resemble the forms of *G. triloba* and *G. cyclostoma*, to which a third type, that of *G. trilocularis* (D'ORBIGNY) might be added. The shape of the latter is approximately that of *G. cyclostoma*, but the shape of the aperture is closer so that of *G. triloba*. If dorsal apertures are lacking or not visible in these immature forms, they are identical with our *Globigerina glutinata*.

In the genus *Globigerina* the adult types are also most distinct, but here there are a lot of small forms to which only doubtful specific names could be given.

Globigerina eggeri RHUMBLER (pl. 5, fig. 1) is the largest *Globigerina* species, of frequent occurrence. It is very distinct. Occasional specimens show *Globoquadrina*-like flaps in the umbilicus.

Globigerina inflata D'ORBIGNY (pl. 5, fig. 2) is much less frequent in our material. The specimens are on the average somewhat smaller than those of *G. eggeri*, of which they are in our material undoubtedly nothing but four-chambered variants. Whether our specimens are really identical with D'ORBIGNY's species is uncertain.

There is little doubt that our **Globigerina subcretacea** LOMNICKI (pl. 4, fig. 23, pl. 5, fig. 5), locally frequent among the middle-sized specimens, is a young stage of *G. eggeri* and *G. inflata* in many instances. It is characterized by a rather small, somewhat depressed early dorsal spiral, so that the individuals approach the *Globigerinella* type. They are closest to LOMNICKI's additional specimens; they are more loosely coiled than BRADY's individuals.

Globigerina radians EGGER (pl. 4, fig. 24, pl. 5, fig. 6) is an occasional, mainly small variant of *G. subcretacea* with about four chambers in the final coil. The chambers are elongated in a radial direction. More nearly three-chambered variants are close to *G. bulbosa* LEROY.

Globigerinella aequilateralis (BRADY) (pl. 5, fig. 4) is not frequent. Some specimens are close to our *Globigerina subcretacea*, but they are mostly distinctly bigger. They are generally more closely coiled than BRADY's types, resembling CUSHMAN's variety *involuta* better.

A species that cannot be clearly linked to any other in our material of the larger Globigerinae, is **Globigerina bulloides** D'ORBIGNY (pl. 4, fig. 16, pl. 5, fig. 3). Such individuals are not frequent; they seem to be of mainly local occurrence. Especially the specimens that reach middle sizes are most typical in apertural features.

Furthermore there are a number of small types that are generally not of very constant size and shape. Partly they may be of local importance only. Commonly they are thin-shelled, composed of numerous chambers. The early chambers are often clearly visible, from which it appears that they have beginnings with very small initial chambers (smaller than 10 μ). It cannot be stated, however, that they are microspheric specimens, belonging to larger individuals of the other species. In larger specimens the test is generally so much thickened that the size of the first chamber cannot be ascertained from the outside, but individuals with similar small first chambers were observed among them as well.

Globigerina bradyi WIESNER (pl. 4, fig. 22, pl. 5, fig. 15) is a small conical type, the specific determination of which was not difficult. It reaches notable frequencies only in some samples (colder water?) off the coast of Paria.

Globigerina cf. **megastoma** EARLAND (pl. 4, fig. 21, pl. 5, fig. 19) is a small type of compact *bulloides* shape (not *bulloides* aperture) with a more elevated dorsal spiral part. The species was originally described for specimens much bigger than ours. Otherwise the scarce individuals resemble the type figures fairly well. Mainly found in a few samples. The specimens resemble *G. bradyi*, but they have a much lower conical shape and are also somewhat bigger.

A few small specimens have been called **Globigerina** cf. **dutertrei** D'ORBIGNY (pl. 4, fig. 20, pl. 5, fig. 14). They have a flat to slightly convex dorsal side, mostly but four chambers in the final whorl, and are characterized by an elongate, oblique aperture. They resemble the figures given by D'ORBIGNY and BRADY. Their relations are obscure.

In our material **Globigerina pachyderma** (EHRENBERG), the typical colder water species, is also rare, if present. Only some very small specimens are close to the type. Others are too thin-walled and with too lobulate periphery. They probably belong to our group of *G.* cf. *dutertrei* and *G.* cf. *megastoma*.

Globigerina quinqueloba NATLAND (pl. 4, fig. 19, pl. 5, fig. 16) could again be identified with more certainty, but here the difficulty is that the variable forms could not be clearly separated during the rapid counting from our *Valvulineria humilis* (BRADY). They are again minute forms.

Globigerina glutinata EGGER (pl. 4, fig. 17, pl. 5, fig. 17) is another very small, distinct form, which has been recognized very often in our material. It resembles *Globigerinoides triloba*, from which it differs in the smaller size, thinner wall and the lack of distinct dorsal apertures. It is not the same form, identified as *Globigerinita glutinata* by PHLEGER (1954–55).

Globigerinita naparimaensis BRONNIMANN (pl. 4, fig. 18, pl. 5, fig. 18) is another small, rare form. Our individuals are identical with those figured by BRONNIMANN from the Miocene of Trinidad and also with those figured as *G. glutinata* (EGGER) (pars) by PHLEGER (1954–55). The basal type of this species seems to be close to our *Globigerina glutinata*. This aberrant form was also encountered among the recent Snellius material from the Indonesian waters. It may represent a similar anomalous form as our *"abnormal" Globigerinoides*.

Finally some rare types, not belonging to *Globigerinoides* and *Globigerina*, have to be mentioned.

Orbulina universa D'ORBIGNY is found in several samples, but it is not frequent.

Pulleniatina obliquiloculata (PARKER and JONES) is another, not frequent species in part of our samples.

Candeina nitida D'ORBIGNY, though present with distinct individuals, is equally rare.

Tretomphalus atlanticus CUSHMAN is also very rare. It does not belong to the family Globigerinidae.

Distribution (including the *Globorotalia* species). The arbitrary limits between the various types and the long duration of the investigation are partly responsible for inconstant determinations. This may be one of the factors that the maps of individual species often showed incomprehensible fluctuations. Sorting may be another important factor. This probably accounts for the greater relative frequency of the larger *Globigerinoides* types in the areas with coarser sediments (map 39).

Most valuable is map 38 with the distribution of the relative totals of all planktonic individuals. With several irregularities it shows a gradual increase of the plankton/benthos ratio from the coast to deeper water. This ratio is 1 or more at several places along the continental slope. Very remarkable are the high ratios immediately north of Paria peninsula. This area contains numerous very small Globigerinidae, with probably cold water forms such as *Globigerina bradyi*. They indicate sedimentation from a colder oceanic current.

Indications of probably older material at the surface are stations with very low relative frequencies. They are mainly found in the deeper part north of the mouth of the Orinoco, such as stations DZ 1173, 1177, DH 1026, 1028 and possibly some others further east.

No relation was found between depth and the percentage of anomalous individuals (*"abnormal" Globigerinoides* and *Globigerinita naparimaensis*) on the total of planktonics. The average is about 10 %, varying irregularly between 0 and 20 %.

Globorotalia

Map 40

Among the planktonic individuals, those of the genus *Globorotalia* are not very frequent. The following species were recognized.

Globorotalia menardii (D'ORBIGNY) is the most common species. Variants identical with *G. tumida* (BRADY) have been found as well.

Globorotalia truncatulinoides (D'ORBIGNY) is present in several of the samples, mainly along the continental slope.

Globorotalia scitula (BRADY) (not D'ORBIGNY) occurs in a small number of samples.

Globorotalia punctulata (D'ORBIGNY) is very variable in our material. Part of the individuals are close to *G. hirsuta*, as figured by BRADY, and by PHLEGER and PARKER (1951). Very rare in our samples.

SYSTEMATIC DESCRIPTION OSTRACODA

Map 41

Over 40 species of Ostracoda were encountered, commonly with low frequencies.

Correct specific, and sometimes even generic determination was not carried out because of the insufficient literature and the minor importance of the individual species. Commonly KEY's determinations for the species of the Gulf of Paria are mentioned as the only reference (1954, Verh. Kon. Ned. Ak. Wet., ser. 1, vol. 20, no. 5). Only the more important species are recorded here.

Distribution. The group as a whole shows a fairly regular distribution on the Orinoco shelf and east of Trinidad, with the higher frequencies (up to 63) in the coarser sediments of the inner part of the shelf, roughly between 35 and 75 metres. West of Tobago higher frequencies reappear near Cumberland Bank. The total numbers of specimens are very small in the pelite areas and along the continental slope.

The counting method, based on benthonic Foraminifera, entails that the observed frequencies of the individual ostracod species are low. As a consequence the distribution patterns are vague and may be incomplete.

For comparison with the data from the Gulf of Paria (KEY, 1954) some 25 species are mentioned below. Open nomenclature is used, since representative material of the various described recent species was not available.

Aurila 1

Pl. 4, fig. 7, 8

Hemicythere aff. *cymbaeformis* (SEGUENZA), KEY, p. 221, pl. 4, fig. 8.

This is one of the rather frequent species with scattered distribution over the entire area. Highest frequencies in the coarse sediments of the outer Orinoco shelf (up to 11).

Bairdoppilata 1

Bairdia amygdaloides BRADY, KEY, p. 219, pl. 4, fig. 1.

Larger specimens, also in KEY's material, clearly show the "hinge" structure of *Bairdoppilata*.

The main distribution of this frequent species (together with rare *Bairdia* spp.) is again in the coarser sediments of the Orinoco shelf and east of Trinidad between 35 and 110 metres. The species obviously avoids the pelite areas. Similar conditions had been found in the Gulf of Paria.

Bradleya 1

Trachyleberis dictyon (BRADY), KEY, p. 222, pl. 4, fig. 9.

Although mainly in the coarser sediments, the distribution is somewhat different from that of the previous species. The higher frequencies are directly outside the pelite area of the Orinoco, not much exceeding the depth of 75 m. This is one of the most common species in our material.

Bythoceratina 1

Bythoceratina aff. *maoria* HORNIBROOK, KEY, p. 226, pl. 5, fig. 9.

Some 12 occurrences. Mainly in the coarser sediments of the outer Orinoco shelf, the area east of Trinidad and Cumberland Bank. Depth range 35–183 m.

Bythoceratina 2

Pl. 4, fig. 9

Some 20 occurrences. Distribution similar to that of *Bythoceratina* 1.

Cypridae spp.

Mostly very small, indeterminable, smooth specimens. Among others, including KEY's *Argilloecia* sp. (p. 218, pl. 3, fig. 8, pl. 6, fig. 1). Probably part of the specimens belong to other families.

Some 40 occurrences with wide depth, sediment and salinity range.

Cytherella 1

Cytherella lata BRADY, KEY, p. 218, pl. 3, fig. 6.

Cytherella 2

Cytherella pulchra BRADY, KEY, p. 218, pl. 3, fig. 7.

For both *Cytherella* species together there are some 12 occurrences, mainly in the shallow pelite and pelitic sediments north of Trinidad and Paria (less than 100 m.). Fewer occurrences on the inner Orinoco shelf with the same depth restriction and often in very pelitic sediment. The species seem to avoid brackish influences.

Cytheropteron 1

Pl. 4, fig. 10

This species has much less ornamentation than our *C.* 2. Occasional specimens seem to be intermediate between these two species, but generally they are well separable.

Cytheropteron 2

Cytheropteron sp. *B* KEY, p. 226, pl. 5, fig. 7.

Cytheropteron 6

Cytheropteron aff. *pyramidale* BRADY, KEY, p. 225, pl. 5, fig. 5.

These three *Cytheropteron* species have a similar distribution in the coarser sediments of the shelf. Highest frequencies occur off the mouth of the Essequibo.

Cytherura 4

Cytherura cf. *lineata* Brady, Key, p. 227, pl. 5, fig. 12.

Possibly more than one species is included. Again mainly in the coarser sediments, but the frequencies are always low.

Eucytherura 1

Eucytherura aff. *complexa* (Brady), Key, p. 228, pl. 5, fig. 13, 14.

Some 20 occurrences in the coarser sediments of the outer Orinoco shelf and Cumberland Bank. Not recorded east of Trinidad, but there are several occurrences around Tobago. Wide depth range.

Eucytherura 2

Pl. 4, fig. 11

Some 15 occurrences, about one third of which is found in the pelite and very pelitic sediment off the Orinoco. Highest frequency of 6 at 35 m. The other occurrences are scattered between 50 and 200 m., partly again in pelite (north of Trinidad).

Hermanites 2

Trachyleberis cf. *thoracophora* (van den Bold), Key, p. 223, pl. 4, fig. 12.

Some 20 occurrences, mainly in the coarser sediments of the Orinoco-eastern Trinidad shelf. Most occurrences between 35 and 75 metres.

Hermanites 5

Bradleya sp. Key, p. 223, pl. 4, fig. 13, pl. 6, fig. 10.

The structure of the hinge and the arrangement of the muscle scars better agree with *Hermanites* than with *Bradleya*.

A fairly frequent species, again in the coarser sediments, mainly in the shallower part of the shelves.

Kangarina 1

Cytheropteron (Kangarina) quellita Coryell and Fields, Key, p. 226, pl. 5, fig. 8.

Some 40 occurrences, always in the coarser sediments. Several occurrences in the shallow area east of Trinidad (20–55 m.), on Cumberland Bank, the outer Orinoco shelf, and the outer part of the Essequibo delta at about 35 m.

Krithe 1

Krithe producta Brady, Key, p. 220, pl. 4, fig. 3, pl. 6, fig. 3.

Some 15 occurrences. North of Trinidad-Paria only in pelite (six occurrences); on the Trinidad-Orinoco shelf in coarser sediments with wide depth range.

Leptocythere 1

Leptocythere aff. *crispata* (BRADY), KEY, p. 220, pl. 4, fig. 5, pl. 6, fig. 5.

Some 15 occurrences in the coarser sediments of the Orinoco shelf. There are occurrences shallower than 40 m. east of Trinidad and in the Essequibo delta. Most others are shallower than 75 m.

Loxoconcha 1

Loxoconcha impressa (BAIRD), KEY, p. 225, pl. 5, fig. 4.

Again mainly in the coarser sediments with low frequencies. At many stations.

Macrocypris 1

Macrocypris decora (BRADY), KEY, p. 219, pl. 3, fig. 10, pl. 6, fig. 4.

Some 15 occurrences with wide depth range.

„Monoceratina" 1

Pl. 4, fig. 12–14

A number of remarkable specimens, as far as we know not belonging to any known genus. Outline, hinge, pores, median sulcus and the long branching hollow spines are shown in the figures. Muscle scars were not visible.

More than 10 occurrences at all depths, all on the Trinidad-Orinoco shelf, none in pelite areas.

Neomonoceratina 1

Paijenborchella (Neomonoceratina) mediterranea RUGGIERI, KEY, p. 228, pl. 5, fig. 15, pl. 6, fig. 12.

Neomonoceratina sp. SWAIN, 1955, Jour .Pal., vol. 29, p.643, pl.64, fig. 15.

Only a few occurrences in shallow water, between 9 and 18 m., off the mouth of the Essequibo.

Paracytheridea 1

Paracytheridea tschoppi VAN DEN BOLD, KEY, p. 220, pl. 4, fig. 4.

Some 20 occurrences, always in coarser sediment, mostly shallower than 80 m.: Cumberland Bank, around Tobago, near eastern Trinidad, and some scattered occurrences on the Orinoco shelf.

Paradoxostoma 1

Paradoxostoma cf. *obliquum* SARS, KEY, p. 229, pl. 5, fig. 17, pl. 6, fig. 14.

Paradoxostoma ensiforme BRADY, SWAIN, 1955, Jour. Pal., vol. 29, p. 633, pl. 63, fig. 7.

Some eight occurrences, most of them shallower than 37 m., but down to 70 m.

Pterygocythereis 1

Pl. 4, fig. 15

Some 15 occurrences, mainly shallower than 80 m. Highest frequencies from 35 to 50 m. in very pelitic sediments.

Xestoleberis 1

Xestoleberis aff. *aurantia* (BAIRD), KEY, p. 228, pl. 5, fig. 16, pl. 6, fig. 13.

At many stations in coarser sediments, mainly those of the outer Orinoco shelf.

SUMMARY

The quantitative study of the Foraminifera in the surface sediment at ± 150 stations on the Orinoco-Trinidad-Paria shelf, was based on a constant number of 200 benthonic individuals per sample. Planktonic species and Ostracoda were counted in addition. The relative frequencies of each species were plotted with contour lines on separate topographic maps. Some 40 of these maps were worth reproducing, because they generally show some relation of the species with the environmental factors: depth, sediment type and salinity.

The distribution of the most frequent species enabled the distinction of a number of faunal zones and areas.

1. The *Rotalia* area in the coastal part of the Orinoco shelf, mainly in silty and sandy pelite, with predominance of *Rotalia beccarii* vars and *R. sarmientoi*, which is probably a consequence of a great tolerance of these species for considerably varying salinity. In part of the outer area, off the Essequibo, the association is considered to be fossil.

2. The *Nonionella-Uvigerina* zone of the inner Orinoco shelf, mainly coinciding with the pelite area. Characteristic species for parts of, or the entire zone, are *Nonionella atlantica*, *Uvigerina peregrina peregrina*, *Virgulina pontoni* and *Cancris sagra*. The distribution of the first species depends mainly on depth, that of the second on pelite, and those of the other two evidently on both these factors. Very slight salinity fluctuations are possibly additional in determining this association.

3. The *Uvigerina* zone in the pelite area north of Trinidad-Paria, characterized by very high percentages of *Uvigerina peregrina peregrina*. Other typical species are *Höglundina elegans*, *Eponides regularis* and *Uvigerina proboscidea*. Fully marine conditions and the pelite evidently determine this type of association. Pelite is considered to be much more fauna-selective than the various gradations of coarser sediment.

4. The shallow reef area east of Trinidad with its coarse sediments, contains an association with many *Amphistegina lessonii* and *Cibicides pseudoungerianus* vars in its seaward part, and with more Miliolidae and arenaceous forms in its shoreward part.

5. The outer part of the Orinoco shelf contains a variable mixture of fossil (Pleistocene) and recent species. The fossil components belong to an association resembling that of the reef area of Trinidad, with the same species. The recent elements are often smaller forms that have their main distribution along the continental slope, depending on a high degree of pelite admixture. The real character of many other species cannot be decided because of the lack of data on living individuals.

The planktonic/benthonic ratio gradually increases in offshore direction. Exception must be made for stations with very high contents of fossil individuals, where the ratios are too low, and for those immediately north of Paria, where they are too high, probably under influence of colder water from the open ocean.

Only the interpretation of the Miocene-Pleistocene faunal succession in the Oranjestad borehole (Aruba) yielded satisfactory results. Generally, published data on fossil faunae contain too much mixture of different associations. For this reason our remarks on the faunae of the Venezuelan Agua Salada group and the Colombian Tubara beds, give but slight new information.

In the descriptive part of the paper remarks on relationships and distribution are only given at some length for the more important species. The Globigerinidae are thought to belong to but a few, highly variable, biological species, which contain several of the forms described as distinct species in the literature. Among the benthonic species but a single one had to be considered as new (*Reussella minuta*); for one other a new name was needed (*Virgulina antilleana*). The encountered Ostracoda contain several new species and even genera, but we refrained from a detailed analysis.

LITERATURE

1. Andel, Tj. van and H. Postma, 1954, Recent sediments of the Gulf of Paria. Verh. Kon. Ned. Ak. Wetensch., afd. Natuurk., ser. 1, vol. 20, no. 5.
2. Bandy, O. L., 1953, Ecology and Paleoecology of some California Foraminifera, Part 1, The frequency distribution of Recent Foraminifera off California; Part 2, Foraminiferal evidence of subsidence rates in the Ventura Basin. Jour. Pal., vol. 27, pp. 161–182 and 200–203.
3. ———, 1954, Distribution of some shallow-water Foraminifera in the Gulf of Mexico. Geol. Survey, Prof. Paper 254–F, pp. 125–140.
4. ———, 1956, Ecology of Foraminifera in Northeastern Gulf of Mexico. Geol. Survey, Prof. Paper 274–G, pp. 179–204.
5. Boltovskoy, E., 1954, Foraminiferos del Golfo San Jorge. Rev. Inst. Nac. Inv. Cien. Nat., Cien. Geol., vol. 3, no. 3, pp. 85–228.
6. ———, 1954, Foraminiferos de la Bahia San Blas. Rev. Inst. Nac. Inv. Cien. Nat., Cien. Geol., vol. 3, no. 4, pp. 247–300.
7. Brady, H. B., 1884, Report Voyage Challenger, Zoology, vol. 9.
8. Crouch, R. W., 1952, Significance of temperature on Foraminifera from deep basins off Southern California coast. Bull. A.A.P.G., vol. 36, pp. 807–843.
9. Cushman, J. A., 1910–1917, A monograph of the Foraminifera of the North Pacific Ocean. U.S. Nat. Mus., Bull. 71.
10. ———, 1918–1931, Foraminifera of the Atlantic Ocean. U.S. Nat. Mus., Bull. 104.
11. ———, 1921, Foraminifera of the Philippine and adjacent seas. U.S. Nat. Mus., Bull. 100, vol. 4.
12. ———, 1922, Shallow-water Foraminifera of the Tortugas region. Carnegie Inst. Wash., Publ. 311.
13. ———, 1926, Recent Foraminifera from Porto Rico. Carnegie Inst. Wash., Publ. 344.
14. ———, 1927, Recent Foraminifera from off the West Coast of America. Bull. Scripps Inst. Oceanogr., Techn. Ser., vol. 1, pp. 119 188.
15. ———, 1934, Smaller Foraminifera from Vitilevu, Fiji. Bernice P. Bishop Mus., Bull. 119, pp. 102–142.
16. ——— and P. Bronnimann, 1948, Additional new species of arenaceous Foraminifera from shallow waters of Trinidad. Contr. Cushm. Lab. Foram. Res., vol. 24, pp. 37–42.
17. Drooger, C. W., 1953, Miocene and Pleistocene Foraminifera from Oranjestad, Aruba (Netherlands Antilles). Contr. Cushm. Found. Foram. Res., vol. 4, pp. 116–147.
18. Ellis, B. F. and A. R. Messina, 1940–1957, Catalogue of Foraminifera.
19. Gould, H. R. and R. H. Stewart, 1956, Continental terrace sediments in the northeastern Gulf of Mexico. Soc. Ec. Pal. Min., Spec. Publ. no. 3, pp. 2–20.
20. Hofker, J., 1956, Foraminifera Dentata; Foraminifera of Santa Cruz and Thatch-Island, Virginia Archipelago, West Indies. Univ. Zool. Mus. København, no. 15.

21. KAASSCHIETER, J. P. H., 1955, The microfauna of the Aquitanian-Burdigalian of southwestern France, Part 3, Smaller Foraminifera. Verh. Kon. Ned. Ak. Wetensch., afd. Natuurk., ser. 1, vol. 21, no. 2, pp. 51–100.
22. KORNFELD, M. M., 1931, Recent littoral Foraminifera from Texas and Louisiana. Contr. Dept. Geol., Stanford Univ., vol. 1, no. 3, pp. 77–93.
23. LOWMAN, S. W., 1949, Sedimentary facies in Gulf Coast. Bull. A.A.P.G., vol. 33, pp. 1939–1997.
24. MARKS, P., 1951, A revision of the smaller Foraminifera from the Miocene of the Vienna Basin. Contr. Cushm. Found. Foram. Res., vol. 2, pp. 33–73.
25. MILLER, D. N., 1953, Ecological Study of the Foraminifera of Mason Inlet, North Carolina. Contr. Cushm. Found. Foram. Res., vol. 4, pp. 41–63.
26. MOORE, D. G., 1955, Rate of deposition shown by relative abundance of Foraminifera. Bull. A.A.P.G., vol. 39, pp. 1594–1600.
27. NATLAND, M. L., 1933, The temperature and depth distribution of some recent and fossil Foraminifera in the southern California region. Bull. Scripps Inst. Oceanogr., Techn. Ser., vol. 3, pp. 225–230.
28. ———, 1938, New species of Foraminifera from off the West Coast of North America and from the later Tertiary of the Los Angeles Basin. Bull. Scripps Inst. Oceanogr., Techn. Ser., vol. 4, pp. 137–164.
29. NORTON, R. D., 1930, Ecologic relations of some Foraminifera. Bull. Scripps Inst. Oceanogr., Techn. Ser., vol. 2, pp. 331–338.
30. PARKER, F. L., 1948, Foraminifera of the continental shelf from the Gulf of Maine to Maryland. Bull. Mus. Comp. Zoology, Harvard Coll., vol. 100, no. 2, pp. 213–241.
31. ———, 1952, Foraminifera species off Portsmouth, New Hampshire. Bull. Mus. Comp. Zoology, Harvard Coll., vol. 106, no. 9, pp. 391–423.
32. ———, 1952, Foraminiferal distribution in the Long Island Sound-Buzzards Bay Area. Bull. Mus. Comp. Zoology, Harvard Coll., vol. 106, no. 10, pp. 425–473.
33. ———, 1954, Distribution of the Foraminifera in the northeastern Gulf of Mexico. Bull. Mus. Comp. Zoology, Harvard Coll., vol. 111, no. 10, pp. 453–588.
34. ———, PHLEGER, F. B. and J. F. PEIRSON, 1953, Ecology of Foraminifera from San Antonio Bay and environs, southwest Texas. Cushm. Found. Foram. Res., Spec. Publ. no. 2.
35. PARKER, R. H., 1956, Macro-invertebrate assemblages as indicators of sedimentary environments in East Mississippi Delta Region. Bull. A.A.P.G., vol. 40, pp. 295–376.
36. PHLEGER, F. B., 1939, Foraminifera of submarine cores from the continental slope. Bull. Geol. Soc. Am., vol. 50, pp. 1395–1422.
37. ———, 1942, Foraminifera of submarine cores from the continental slope, part 2. Bull. Geol. Soc. Am., vol. 53, pp. 1073–1098.
38. ———, 1952, Foraminifera Ecology off Portsmouth, New Hampshire. Bull. Mus. Comp. Zoology, Harvard Coll., vol. 106, no. 8, pp. 318–390.
39. ———, 1954–1955, Foraminifera and deep-sea research. Deep-sea Research, vol. 2, pp. 1–23.
40. ———, 1955, Ecology of Foraminifera in southeastern Mississippi delta area. Bull. A.A.P.G., vol. 39, pp. 712–752.
41. ———, 1956, Significance of living foraminiferal populations along the Central Texas Coast. Contr. Cushm. Found. Foram. Res., vol. 7, pp. 106–151.
42. ——— and F. L. PARKER, 1951, Ecology of Foraminifera northwest Gulf of Mexico. Geol. Soc. Am., Mem. 46.

43. REDMOND, C. D., 1953, Miocene Foraminifera from the Tubara beds of northern Colombia. Jour. Pal., vol. 27, pp. 708–733.

44. RENZ, H. H., 1948, Stratigraphy and fauna of the Agua Salada Group, State of Falcón, Venezuela. Geol. Soc. Am., Mem. 32.

45. SHEPARD, F. P., 1956, Marginal sediments of Mississippi delta. Bull. A.A.P.G., vol. 40, pp. 2537–2623.

46. ——— and D. G. MOORE, 1955, Central Texas Coast sedimentation: Characteristics of sedimentary environment, recent history, and diagenesis. Bull. A.A.P.G., vol. 39, pp. 1463–1593.

47. SWAIN, F. M., 1955, Ostracoda of San Antonio Bay, Texas. Jour. Pal., vol. 29, pp. 561–646.

48. WALTON, W. R., 1955, Ecology of living benthonic Foraminifera, Todos Santos Bay, Baja California. Jour. Pal., vol. 29, pp. 952–1108.

49. WESTERMANN, J. H., 1951, The water bore of Oranjestad 1942–1943, and its implication as to the geology and geohydrology of the island of Aruba (Netherlands West Indies), I and II. Proc. Kon. Ned. Ak. Wetensch., ser. B. vol. 54, pp. 140–159.

50. TODD, R. and P. BRONNIMANN, 1957, Recent Foraminifera and Thecamoebina from the eastern Gulf of Paria. Cushm. Found. Foram. Res., Spec. Publ. no. 3.

51. KOLDEWIJN, B. W., 1958, Sediments of the Paria-Trinidad shelf. Rep. Orinoco shelf Exped., vol. 3. Mouton & Co, The Hague.

52. NOTA, D. J. G., 1958, Sediments of the Western Guiana shelf. Rep. Orinoco shelf Exped., vol. 2. H. Veenman & Zonen, Wageningen

EXPLANATION OF PLATE 1

Figs.		Page
1.	*Bifarina advena* (CUSHMAN). (DH 1041). × 16	26
2a, b.	*Bolivina barbata* PHLEGER and PARKER. 2a, side view; 2b, apertural view (ED 1203). × 75.	28
3a, b.	*Bolivina goësi* CUSHMAN. 3a, side view; 3b, apertural view (DY 1169). × 75	28
4a, b.	*Bolivina marginata* CUSHMAN. 4a, side view; 4b, apertural view (DI 1046). × 110	29
5a, b.	*Bolivina paula* CUSHMAN and PONTON. 5a, side view; 5b, apertural view (DB 1003). × 110	30
6a, b.	*Bolivina striatula* CUSHMAN var. *fragilis* PHLEGER and PARKER. 6a, side view; 6b, apertural view (DV 1149). × 70	31
7a, b.	*Bolivina subspinescens* CUSHMAN. 7a, side view; 7b, apertural view (DO 1086). × 110.	32
8a-c.	*Bulimina marginata* D'ORBIGNY. 8a, 8b, opposite views; 8c, apertural view (DI 1046). × 75	33
9a, b.	*Buliminella elegantissima* (D'ORBIGNY). 9a, 9b, opposite views (DV 1144). × 110	34
10a, b.	*Carpenteria proteiformis* GOËS. 10a, side view; 10b, basal view (DZ 1173). × 8	35
11a-c.	*Cassidulina laevigata* D'ORBIGNY. 11a, 11b, side views; 11c, apertural view (ED 1203). × 75.	35
12a-c.	*Cassidulina neocarinata* THALMANN. 12a, 12b, side views; 12c, apertural view (DH 1037). × 75	36
13a-c.	*Cassidulina neocarinata* THALMANN var. 13a, 13b, side views; 13c, apertural view (DH 1041). × 110	37
14a-c.	*Cassidulina norcrossi* CUSHMAN subsp. *australis* PHLEGER and PARKER. 14a, 14b, side views; 14c, apertural view (DH 1047). × 75. . . .	37
15a-c.	*Cassidulina palmerae* BERMUDEZ and ACOSTA. 15a, 15b, side views; 15c, apertural view (EE 1210). × 75	37
16a-c.	*Cassidulina subglobosa* BRADY. Big specimen. 16a, 16b, side views; 16c, apertural view (291). × 50	37
17a-c.	*Cassidulina subglobosa* BRADY var. *subcalifornica* DROOGER. Thick specimen. 17a, 17b, side views; 17c, apertural view (ED 1202). × 75	38

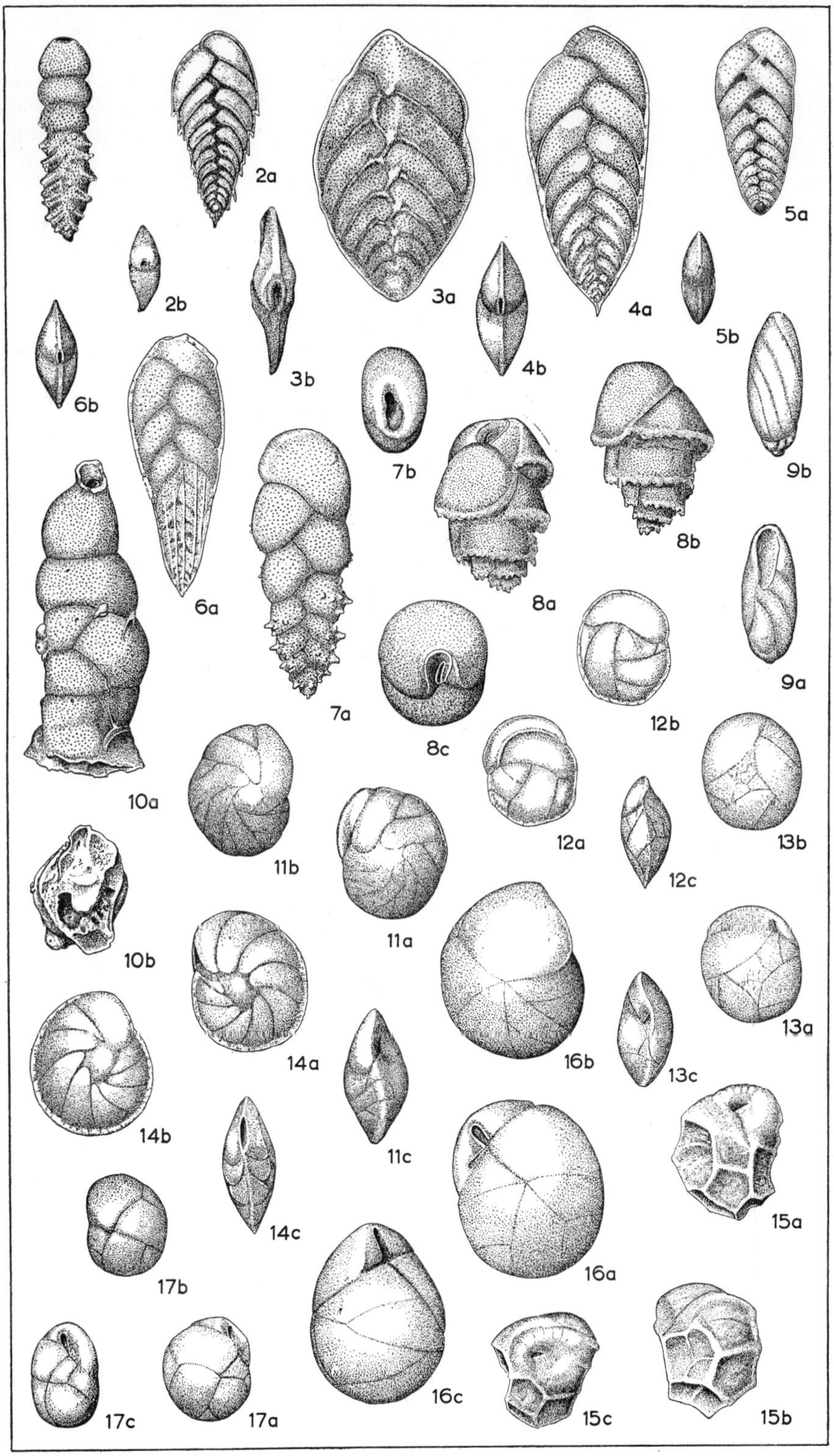
2a
5a
2b
3a
4a
5b
6b
3b
4b
7b
9b
8b
6a
8a
9a
7a
12b
8c
10a
12a
13b
11b
12c
10b
11a
13a
14a
16b
13c
14b
11c
15a
14c
17b
16a
16c
17c
17a
15c
15b

PLATE I

Explanation of plate 2

Figs.		Page
1a-c.	*Cibicides pseudoungerianus* (Cushman). lobulate variant. 1a, ventral view; 1b, dorsal view; 1c, peripheral view. (DO 1087). × 50 . . .	39
2a-c.	*Cibicides pseudoungerianus* (Cushman) var. *antilleana* Drooger. 2a, ventral view; 2b, dorsal view; 2c, peripheral view (DZ 1171). × 50	39,40
3a-c.	*Cibicides pseudoungerianus* (Cushman) var. *robertsonianus* (Brady). 3a, ventral view; 3b, dorsal view; 3c, peripheral view. (EA 1182). × 50	39,41
4a-c.	*Cibicides pseudoungerianus* (Cushman) var. *io* Cushman. 4a, ventral view; 4b, dorsal view; 4c, peripheral view. (DW 1153). × 50 . . .	39,40
5a-c.	*Discorbis bulbosa* Parker. 5a, ventral view; 5b, dorsal view; 5c, peripheral view. (DH 1041). × 75	41
6a-c.	*Discorbis floridanus* Cushman. 6a, ventral view; 6b, dorsal view; 6c, peripheral view. (ED 1202). × 75	42
7a-c.	*Ehrenbergina spinea* Cushman. 7a, ventral view; 7b, dorsal view; 7c, peripheral view. (DW 1153). × 75	43
8a-c.	*Eponides repandus* (Fichtel and Moll). 8a, ventral view; 8b, dorsal view; 8c, peripheral view (DB 1005). × 32	46
9a-c.	*Eponides repandus* (Fichtel and Moll) cf. *Poroeponides lateralis* (Terquem) Phleger and Parker. 9a, dorsal view; 9b, ventral view; 9c, peripheral view (DH 1041). × 50	46
10a, b.	*Gaudryina aequa* Cushman. 10a, side view; 10b, apertural view (DY 1168). × 75	47
11a, b.	*Gaudryina aequa* Cushman. 11a, side view; 11b, apertural view (DB 1000). × 75	47
12a, b.	*Gaudryina atlantica* (Bailey). 12a, side view; 12b, apertural view (DH 1037). × 16	47

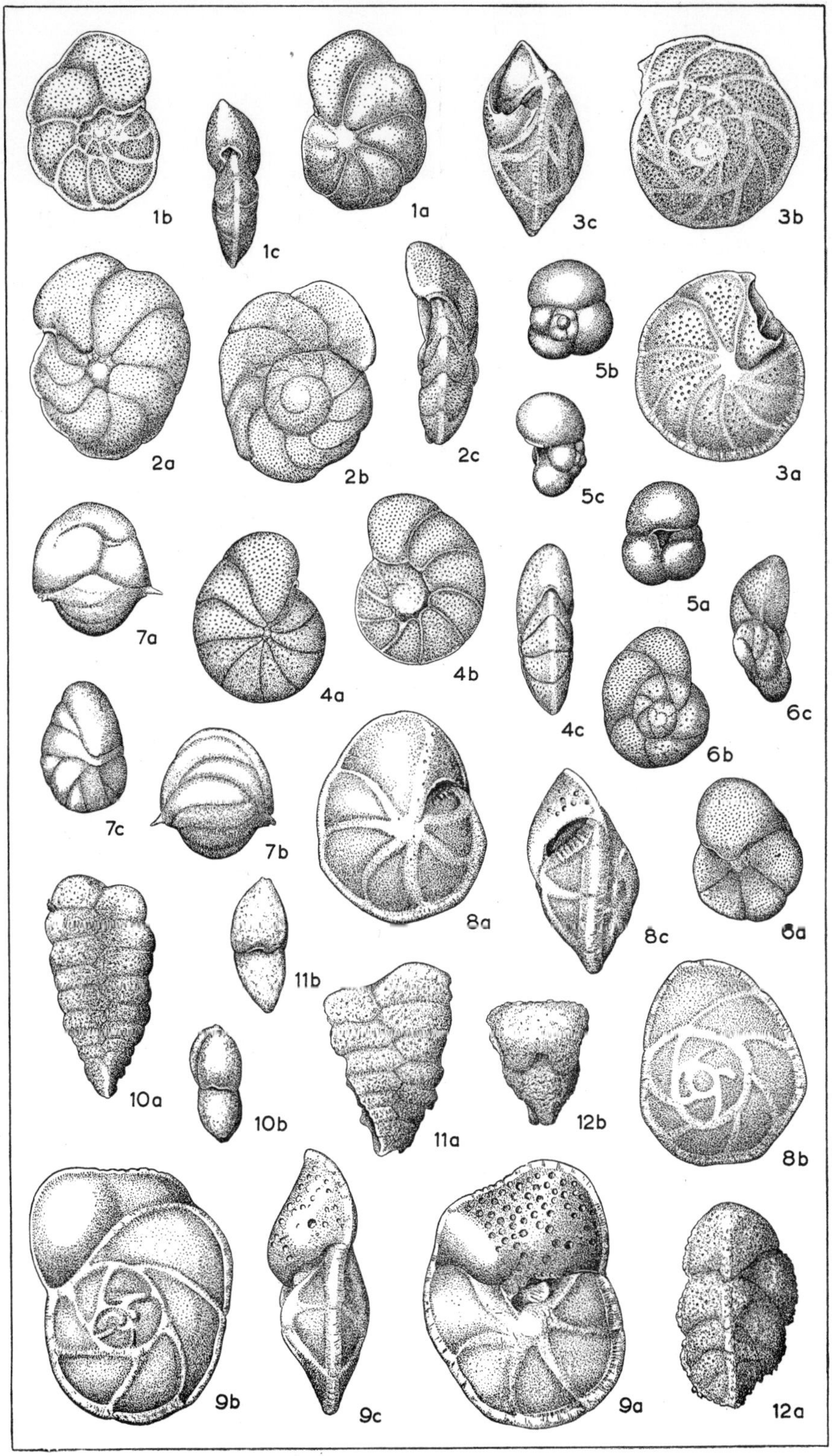
1b
1c
1a
3c
3b
2a
2b
2c
5b
5c
3a
7a
4a
4b
5a
4c
6c
6b
7c
7b
8a
8c
8a
11b
10a
10b
11a
12b
8b
9b
9c
9a
12a

PLATE II

Explanation of plate 3

Figs.		Page
1a, b.	*Gaudryina quadrangularis* Bagg var. *antillana* Bermudez and Acosta. 1a, side view; 1b, apertural view (DI 1047). × 32	47
2a-c.	*Gyroidina* cf. *neosoldanii* Brotzen. 2a, ventral view; 2b, dorsal view; 2c, peripheral view (DB 1006). × 110	48
3a, b.	*Karreriella bradyi* (Cushman). 3a, side view; 3b, apertural view (DZ 1172). × 50	50
4a, b.	*Loxostomum limbatum* (Brady). 4a, side view; 4b, apertural view (EB 1193). × 50	53
5a, b.	*Marginulina subaculeata* (Cushman) var. *glabrata* (Cushman). 5a, side view; 5b, apertural view (DH 1037). × 50	54
6a-c.	*Miliolinella fichteliana* (d'Orbigny). 6a, 6b, lateral views; 6c, apertural view (EB 1193). × 75.	54
7.	*Nodosaria intercellularis* Brady (DB 1003). × 32	55
8.	*Pavonina atlantica* Cushman (EA 1185). × 16	58
9.	*Pseudoclavulina mexicana* (Cushman) (DZ 1172). × 16	60
10a-c.	*Reussella minuta* Drooger and Kaasschieter n. sp. Holotype. 10a, 10b, side views; 10c, apertural view (DB 1001). × 110	64
11a-c.	*Siphonina reticulata* (Czjzek). 11a, ventral view; 11b, dorsal view; 11c, peripheral view (ED 1202). × 75	71
12a-c.	*Sphaeroidina bulloides* d'Orbigny. 12a, apertural view; 12b, 12c, side views (DE 1020). × 75	72
13a, b.	*Textularia foliacea* Heron-Allen and Earland var. *occidentalis* Cushman. 13a, side view; 13b, apertural view (DB 1002). × 50 .	75
14a, b.	*Textularia foliacea* Heron-Allen and Earland var. *occidentalis* Cushman. 14a, side view; 14b, apertural view. (DY 1167). × 75 . .	75
15a, b.	*Textularia mexicana* Cushman. 15a, side view; 15b, apertural view (DH 1037) × 32	75

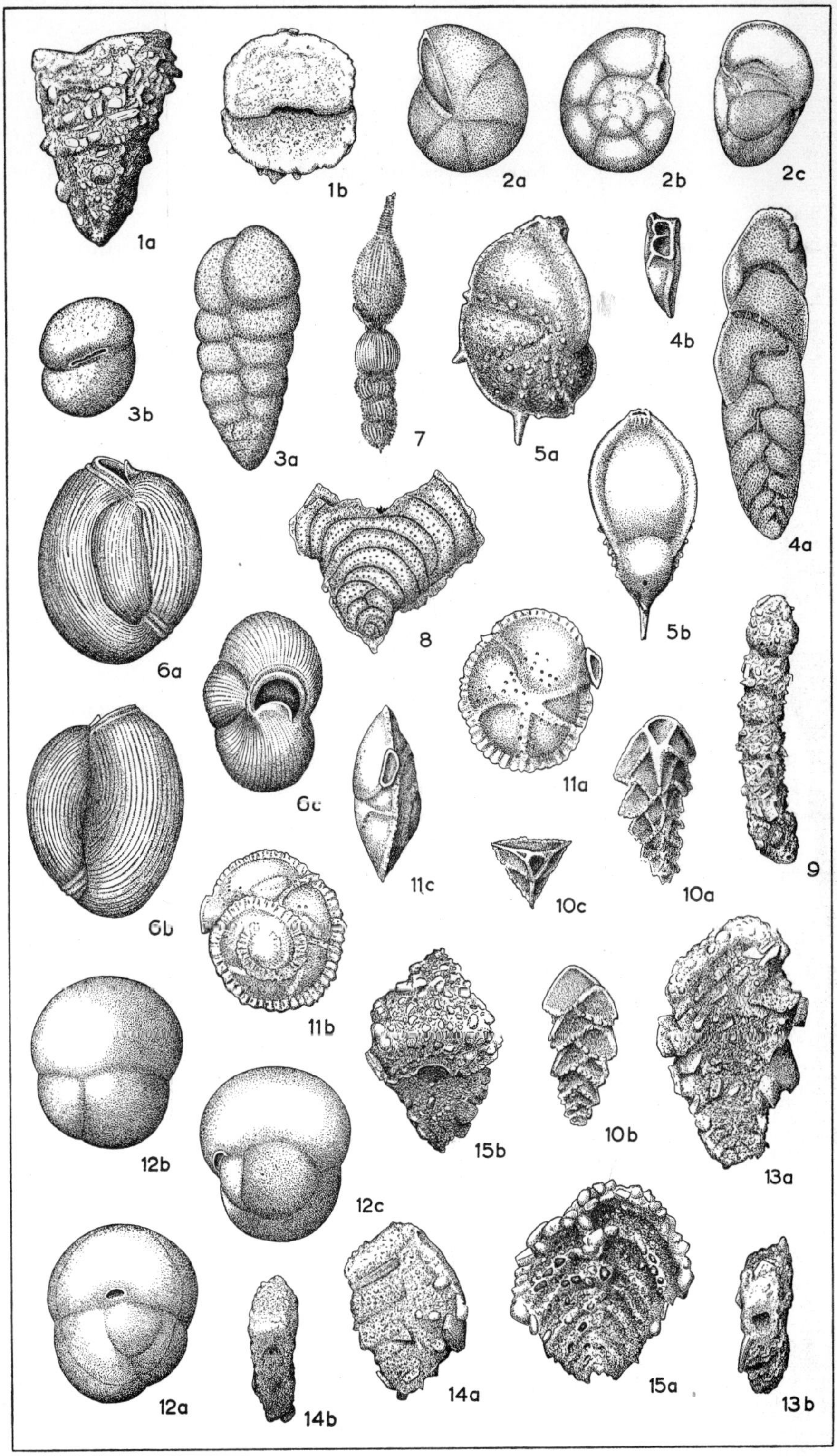
1a
1b
2a
2b
2c
3a
3b
4a
4b
5a
5b
6a
6b
6c
7
8
9
10a
10b
10c
11a
11b
11c
12a
12b
12c
13a
13b
14a
14b
15a
15b

PLATE III

Explanation of plate 4

Figs.		Page
1a, b.	*Textulariella barrettii* (Jones and Parker). 1a, side view; 1b, apertural view (DW 1156). × 8	75
2.	*Uvigerina flinti* Cushman (294). × 50	76
3.	*Vaginulina planata* (Phleger and Parker) (DG 1026). × 16	80
4.	*Virgulina antilleana* Drooger and Kaasschieter nom. nov. (DH 1041). × 110	81
5.	*Virgulina complanata* Egger (DU 1142). × 110	81
6.	*Virgulina spinicostata* Phleger and Parker (DH 1039). × 110	82
7, 8.	*Aurila* 1. 7, lateral view left valve (DR 1108). × 50; 8, internal view right valve (DV 1149). × 50	88
9.	*Bythoceratina* 2. lateral view right valve (DB 1000). × 75	89
10.	*Cytheropteron* 1. lateral view left valve (ED 1202). × 75	89
11.	*Eucytherura* 2. lateral view left valve (DH 1037). × 75	90
12a, b, 13, 14.	"*Monoceratina*", 1. 12a, lateral view left valve; 12b, dorsal view left valve (EA 1183). × 75; 13, internal view left valve (EA 1183). × 70; 14, hinge of right valve (DG 1029). × 50	91
15.	*Pterygocythereis* 1. lateral view left valve (DM 1075). × 75	92
16a-c.	*Globigerina bulloides* d'Orbigny. 16a, ventral view; 16b, dorsal view; 16c, peripheral view (DE 1016). × 75	84
17a-c.	*Globigerina glutinata* Egger. 17a, ventral view; 17b, dorsal view; 17c, peripheral view (DH 1037). × 75	85
18a-c.	*Globigerinita naparimaensis* Bronnimann. 18a, ventral view; 18b, dorsal view; 18c, peripheral view (DY 1168). × 75	85
19a-c.	*Globigerina quinqueloba* Natland. 19a, ventral view; 19b, dorsal view; 19c, peripheral view (DY 1168). × 75	85
20a-c.	*Globigerina* cf. *dutertrei* d'Orbigny. 20a, ventral view; 20b, dorsal view; 20c, peripheral view (DH 1037). × 75	85
21a-c.	*Globigerina* cf. *megastoma* Earland. 21a, ventral view; 21b, dorsal view; 21c, peripheral view (DH 1039). × 75	85
22a-c.	*Globigerina bradyi* Wiesner. 22a, ventral view; 22b, dorsal view; 22c, peripheral view (DB 1010). × 75	85
23a-c.	*Globigerina subcretacea* Lomnicki. 23a, ventral view; 23b, dorsal view; 23c, peripheral view (DY 1168). × 75	84
24a-c.	*Globigerina radians* Egger. 24a, ventral view; 24b, dorsal view; 24c, peripheral view (DV 1150). × 75	84

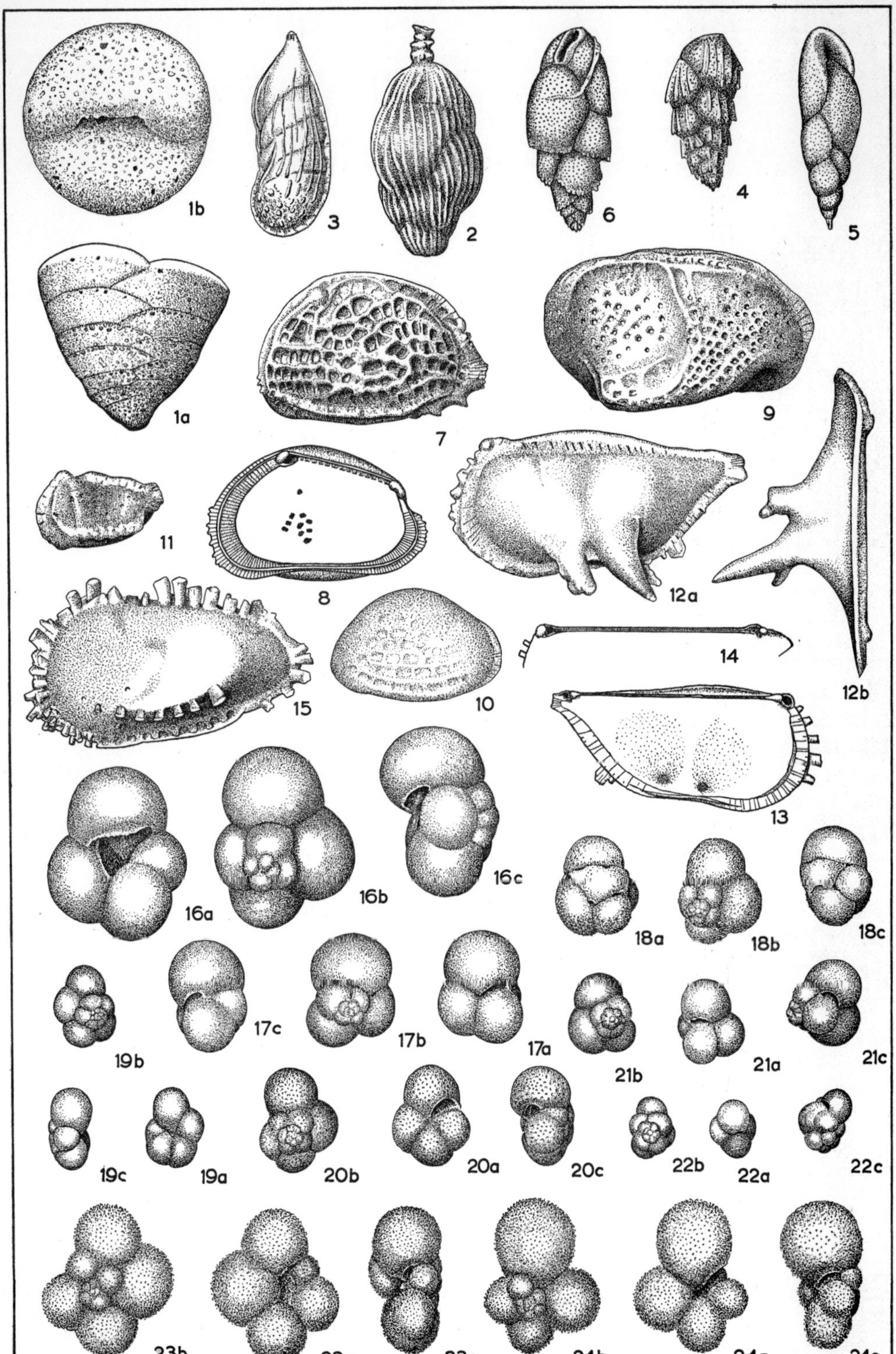
1b
3
2
6
4
5
1a
7
9
11
8
12a
12b
15
10
14
13
16a
16b
16c
18a
18b
18c
19b
17c
17b
17a
21b
21a
21c
19c
19a
20b
20a
20c
22b
22a
22c
23b
23a
23c
24b
24a
24c

PLATE IV

EXPLANATION OF PLATE 5

All figures × 32.

Figs.		Page
1a-c.	*Globigerina eggeri* RHUMBLER. 1a, ventral view; 1b, dorsal view; 1c, peripheral view (DI 1055)	84
2a-c.	*Globigerina inflata* D'ORBIGNY. 2a, ventral view; 2b, dorsal view; 2c, peripheral view (DR 1108)	84
3a-c.	*Globigerina bulloides* D'ORBIGNY. 3a, ventral view; 3b, dorsal view; 3c, peripheral view (DE 1016)	84
4a-c.	*Globigerinella aequilateralis* (BRADY). 4a, 4b, opposite side views; 4c, peripheral view. (291)	84
5a-c.	*Globigerina subcretacea* LOMNICKI. 5a, ventral view; 5b, dorsal view; 5c, peripheral view (DY 1168)	84
6a-c.	*Globigerina radians* EGGER. 6a, ventral view; 6b, dorsal view; 6c, peripheral view (DV 1150)	84
7a-c.	*Globigerinoides cyclostoma* (GALLOWAY and WISSLER). 7a, ventral view; 7b, dorsal view; 7c, peripheral view (DH 1037)	82
8a-c.	*Globigerinoides triloba* (REUSS). 8a, ventral view; 8b, dorsal view; 8c, peripheral view (DH 1037)	83
9a-c.	*Globigerinoides sacculifera* (BRADY). 9a, ventral view; 9b, dorsal view; 9c, peripheral view (DO 1086)	83
10a-c.	*Globigerinoides rubra* (D'ORBIGNY). 10a, ventral view; 10b, dorsal view; 10c, peripheral view (DH 1037)	83
11a-c.	"Abnormal" *Globigerinoides*. 11a, ventral view; 11b, dorsal view; 11c, peripheral view (DY 1168)	83
12a-c.	*Globigerinoides conglobata* (BRADY). 12a, ventral view; 12b, dorsal view; 12c, peripheral view (DZ 1174)	83
13a-c.	*Sphaeroidinella dehiscens* (PARKER and JONES). 13a, ventral view; 13b, dorsal view; 13c, peripheral view (DY 1168)	83
14a-c.	*Globigerina* cf. *dutertrei* D'ORBIGNY. 14a, ventral view; 14b, dorsal view; 14c, peripheral view (DH 1037)	85
15a-c.	*Globigerina bradyi* WIESNER. 15a, ventral view; 15b, dorsal view; 15c, peripheral view (DB 1010)	85
16a-c.	*Globigerina quinqueloba* NATLAND. 16a, ventral view; 16b, dorsal view; 16c, peripheral view (DY 1168)	85
17a-c.	*Globigerina glutinata* EGGER. 17a, ventral view; 17b, dorsal view; 17c, peripheral view (DH 1037)	85
18a-c.	*Globigerinita naparimaensis* BRONNIMANN. 18a, ventral view; 18b, dorsal view; 18c, peripheral view (DY 1168)	85
19a-c.	*Globigerina* cf. *megastoma* EARLAND. 19a, ventral view; 19b, dorsal view; 19c, peripheral view (DH 1039)	85

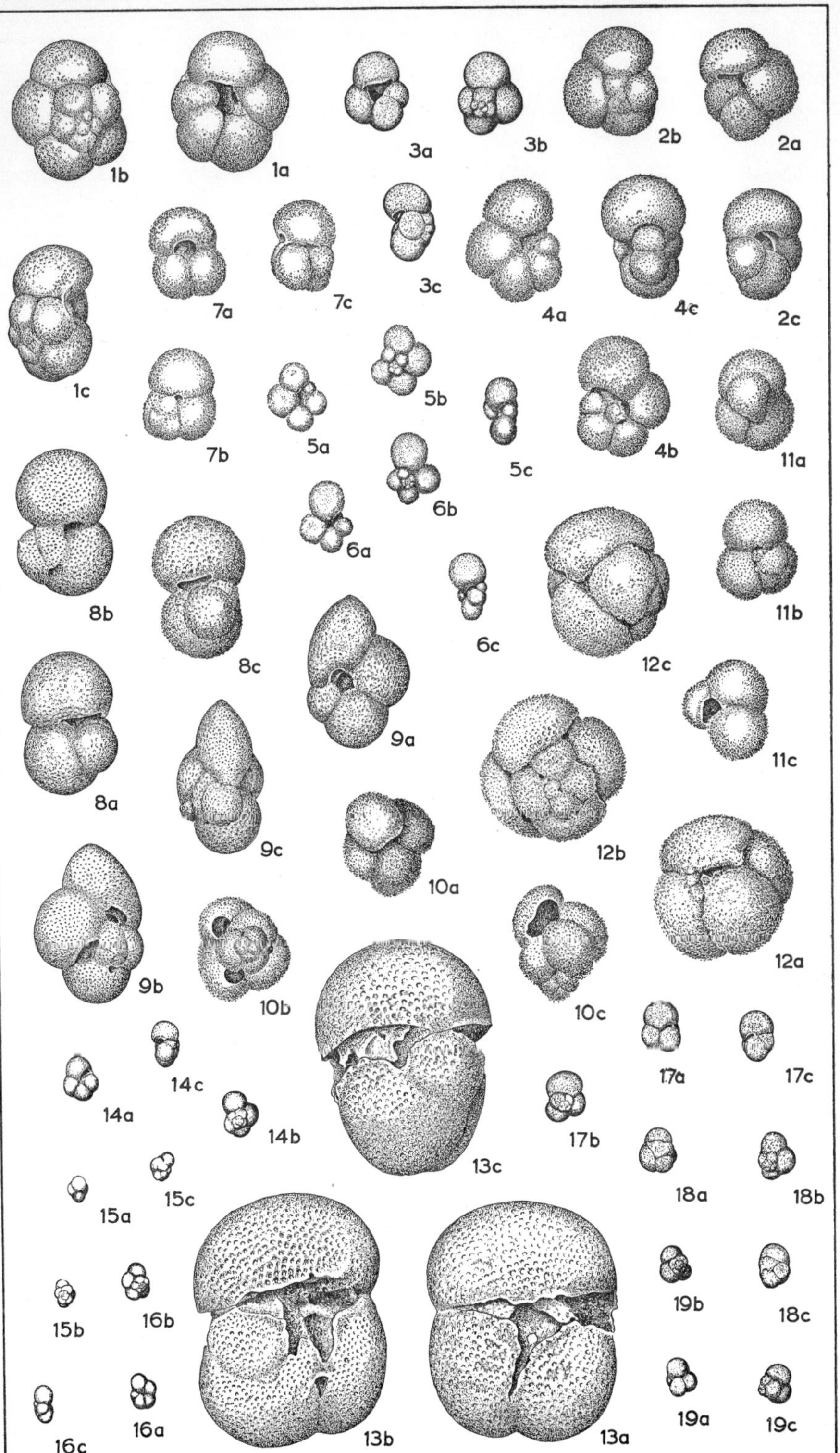
1b
1a
3a
3b
2b
2a
7a
7c
3c
4a
4c
2c
1c
7b
5a
5b
5c
4b
11a
8b
6b
6a
11b
8c
6c
12c
9a
11c
8a
9c
10a
12b
9b
10b
10c
12a
14c
14a
14b
13c
17a
17c
17b
15c
15a
18a
18b
15b
16b
19b
18c
16c
16a
13b
13a
19a
19c

PLATE V

EXPLANATION OF THE MAPS

Depth lines are at intervals of 20 fathoms. As a consequence of this wide spacing, some topographic features, such as the gully north of the Essequibo, are not apparent.

Station numbers and geographic names are to be found on map 1, the distribution of sediments on map 2.

If two species, variants or growth stages have been entered separately on the same map, the quantities of one have been underlined. The contour lines have been drawn for the totals.

List of samples, in which less than 200 benthonic individuals could be counted.

Sample number	Number of specimens
DL 1068	153
DK 1066	26
DK 1065	21
DK 1064	50
DI 1052	70
DI 1051	50
DZ 1173	178
DO 1092	95
DP 1094	15
DP 1095	150
DH 1045	5
DH 1044	32
DH 1042	155
DP 1096	52
DP 1098	39
DG 1036	132
DG 1026	142

On the maps these quantities have been converted to 200. They have been placed in circles.

ERRATA

On some of the maps the contour lines may have shifted slightly with respect to the topographic background.

Occasional contour lines have fallen out (maps 3, 7, 12) or have been given a wrong number (maps 12, 17).

On map 26 the station number of the figured *Reussella atlantica* is DH 1040.

On map 33 both sets of figures must be shifted. The figures of *Textularia candiana* now accompany the text of *T. conica*, and reciprocally.